CLIMATE OF EXTREMES

PATRICK J. MICHAELS and ROBERT C. BALLING JR

CLIMATE OF EXTREMES

Global Warming Science They Don't Want You to Know

CATO
INSTITUTE
WASHINGTON, D.C.

Second printing: July 2009

First paperback printing: 2010

Library of Congress Cataloging-in-Publication Data

Michaels, Patrick J.
Climate of extremes : global warming science they don't want you to know / Patrick J. Michaels and Robert C. Balling Jr.
p. cm.
Includes bibliographical references and index.
ISBN-13: 978-1-933995-23-6 (alk. paper)
ISBN-10: 1-933995-23-8 (alk. paper)
1. Global warming. 2. Climatic changes. I. Balling, Robert C. II. Title.

QC981.8.G56M534 2008
551.6--dc22 2008052342

Printed in the United States of America.

CATO INSTITUTE
1000 Massachusetts Ave., N.W.
Washington, D.C. 20001
www.cato.org

Contents

Epigraph

"We have 25 years or so invested in the work. Why should I make the data available to you, when your aim is to try and find something wrong with it?"

—Phil Jones, developer of the United Nations Intergovernmental Panel on Climate Change temperature history, in a letter to Australian climatologist Warrick Hughes, February 21, 2005

Preface

At the end of June 2009, I will be leaving the University of Virginia, as fine a public school as there is in the world. The university cannot guarantee me both academic freedom and a full salary from the Commonwealth of Virginia. My faculty position was "Research Professor and State Climatologist, Department of Environmental Sciences." My salary was paid in its large majority by a separate line in the university's budget, labeled "State Climatology Office," itself a part of the overall budget for the Commonwealth of Virginia.

I was appointed Virginia State Climatologist on July 7, 1980. Like most other State Climatologists, I was faculty at a major public institution, and the appointment was without term, although the faculty position itself was without academic tenure. It was nonetheless subject to the same review process (without teaching duties) for promotion to associate and then to full professor.

I served Republican and Democratic administrations. I met all the Virginia governors. I really liked Republican Governor George Allen. I told Governor Jim Gilmore, also a Republican, how fortunate I was to be able to speak the truth on climate change, even as it was becoming politically unpopular. I was incredibly impressed by the professional staff that served Democrat Mark Warner. His staff members were as good as or better than many federal staffers I have worked with.

Given the political nature of climate change, it was only a matter of time until some governor went after his State Climatologist. I'll be happy to say I brought it on myself. I'm articulate, chatty, and, thanks to the Cato Institute, have great access to TV, radio, and major news outlets. I fully used my privileges as a University of Virginia faculty member, which included the right to consult for whomever I wanted without jeopardizing my position or the academic freedom that went with it.

Which meant, of course, consulting for entities ranging from the Environmental Protection Agency to power producers with a dog in the global warming hunt. One of those was Intermountain Rural Electric Association, a small Colorado utility. When my work for them became public knowledge, Virginia Governor Timothy Kaine told me not to speak as State Climatologist when it came to global warming. If the State Climatologist is a political appointment, that's his call. If it is a lifetime honorific, it's not. But regardless of which of those it is, almost all my university salary was contingent upon my being State Climatologist.

The University of Virginia valiantly, if clumsily, attempted to paper this over. All of a sudden, I was told I should no longer refer to myself as Virginia State Climatologist. Instead, I should cite my seal of certification as Director of the Virginia State Climatology Office, given by the American Association of State Climatologists (AASC). The position of State Climatologist had apparently become a political appointment.

I wasn't asked to do the impossible, merely the impossibly awkward. The University of Virginia Provost wrote to me:

> You should refer to yourself as the "AASC-designated state climatologist" and your office as the "AASC-designated State Climatology Office," or if you prefer, "AASC-designated State Climatology Office at the University of Virginia." I recognize that the titles may be awkward but the message from the Governor's Office was very clear about what they expected.

Needless to say, this quickly became unworkable. Newspaper editors wouldn't suffer such encumbering verbiage, it didn't fit on a TV Chiron, and making a disclaimer every time I spoke about climate that my views didn't reflect those of the Commonwealth of Virginia or the University of Virginia (despite their being correct!) would never fit in a sound bite. So I had the choice of speaking on global warming and having my salary line terminated, or leaving.

Other State Climatologists soon had similar difficulties. George Taylor at Oregon State University, who is very popular with the AASC (and the only person ever elected to consecutive terms as president), was told that he was simply not to speak on global warming. Having read the playbook established by Governor Kaine in Virginia, Governor Ted Kulongoski (D) told Portland's KGW-TV that "Taylor's contradictions interfere with the state's stated goals to reduce greenhouse gases."

Taylor had long questioned glib statements about a 50 percent decline in Pacific Northwest snowpack, which were being made by climate alarmists worldwide. The 50 percent figure is only part of the story. That figure accrues if one starts the data in 1950 and ends in the mid-1990s. If one uses the entire set of snowpack data (1915–2004), a different picture emerges (Figure P. 1, bottom). Taylor was told to shut up as State Climatologist even though he was merely telling the truth.

Taylor resigned his Oregon State University position in February 2008.

David Legates, at the University of Delaware, was told by Governor Ruth Ann Minner (D) that he could no longer speak on global warming as State Climatologist. His faculty position is a regular tenured line in the geography department. He's free, as State Climatologist, to say anything about the weather, so long as there's no political implication. Unfortunately, as most State Climatologists will attest, most reporters specifically ask whether this or that unusual storm or unusually hot (or cold!) day is related to global warming. Scientists who refuse to answer that question don't get return calls.

Minner was upset because Legates was an author of an amicus brief to the U.S. Supreme Court (Baliunas et al.) in its first global warming–related case, *Massachusetts v. U.S. Environmental Protection Agency.* Baliunas et al. sided with the federal government (namely the Environmental Protection Agency [EPA]), which maintained that it was not required to issue regulations reducing carbon dioxide emissions. Justice Antonin Scalia cited Baliunas et al. in his dissent, as the court voted 5-4 that it was within the EPA's purview to propose and then enforce carbon dioxide limitations.

So Legates stopped speaking about global warming as Delaware's State Climatologist.

Out West, things got even uglier. The Assistant State Climatologist for Washington, Mark Albright, was fired because, despite his boss's orders, he refused to stop e-mailing—to journalists, to inquiring citizens, to *anyone*—the entire snowfall record for the Cascade Mountains rather than the cherry-picked one. For e-mailing that record, the assistant state climatologist in Washington lost his job.

What had started with Oregon's George Taylor had migrated across the Columbia River.

Figure P. 1
SNOWPACK IN MAXIMUM WINTER MONTH, EXPRESSED AS DEPARTURE FROM THE 1971–2000 AVERAGE: FOR 1950–2004 (TOP) AND FOR 1915–2004 (BOTTOM)

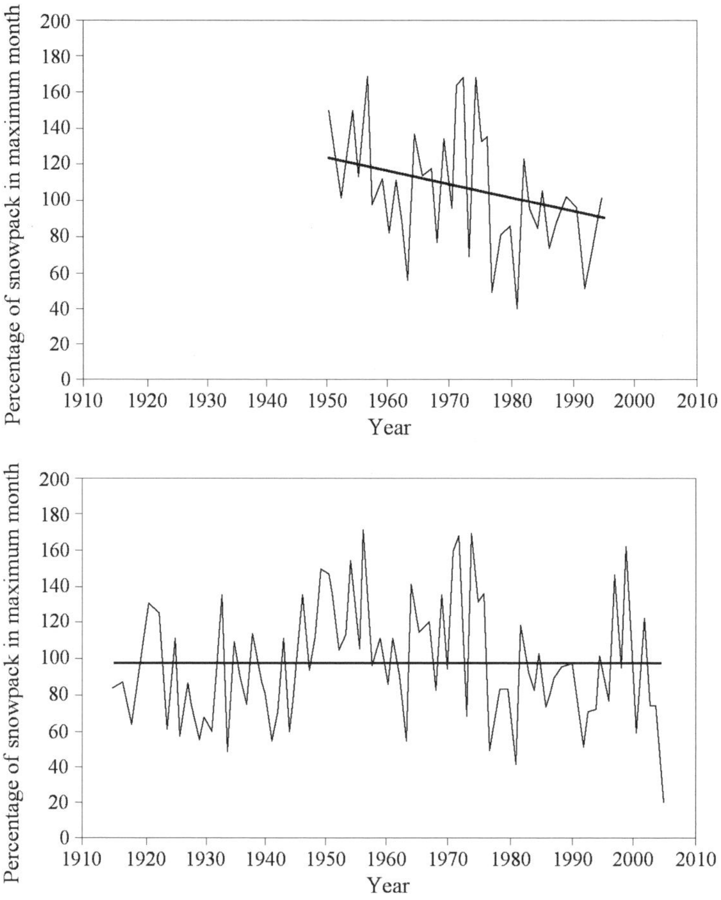

SOURCE: Jones 2007.

NOTE: Using all the data back to 1915 clearly shows that the current era is hardly unusual, despite one very low reading in 2004.

State Climatologist Phil Mote terminated Albright. Both positions were in the University of Washington's atmospheric science department, one of the world's best. A senior member of that department,

Professor Clifford Mass, commented, "In all my years of doing science, I've never seen this sort of gag-order approach to doing science."

What is so scary that some governors don't want you to know it?

Apparently it is this: The world is not coming to an end because of global warming. Further, we don't really have the means to significantly alter the temperature trajectory of the planet. All of this will be spelled out in considerable detail within the rest of this book.

Governors Kaine, Kulongoski, and Minner, this book's for you!

We would like to acknowledge the considerable effort put into the research for this manuscript by Chip Knappenberger and Robert E. Davis. Peter VanDoren, David Boaz, Sonja Boehmer-Christiansen provided invaluable review comments. Rola Brentlin and Jonathan Eidsness also provided insightful reviews. Amy Lemley cheerfully did some extensive copyediting, making a boring global warming story into something readable and, maybe, enjoyable. Thanks to all of you for all your help.

—Patrick J. Michaels
Washington, DC, September 2008

Foreword: A Climate of Extremes

Something about the global warming debate has changed, and changed for the worse. The debate itself has become a climate of extremes. Truth and fact no longer matter, outrageous exaggerations go unchallenged, unscientific speculation is unquestionably accepted, and nonbelievers lose their jobs.

Consider this interview with former Vice President Al Gore, on *Larry King Live*, May 22, 2007:

> UNIDENTIFIED WOMAN: Vice President Al Gore, what issues caused by climate change globally are likely to affect the United States security in the next 10 years?
>
> GORE: You know, even a one-meter increase, even a three-foot increase in sea level would cause tens of millions of climate refugees.
>
> If Greenland were to break up and slip into the sea or West Antarctica, or half of either and half of both, it would be a 20-feet increase, and that would lead to more than 450 million climate refugees.
>
> The direct impacts on the U.S. have already begun. Today, 49 percent of America is in conditions of drought or near-drought. And we have had droughts in the past, but the odds of serious droughts increase when the average temperatures go up, as they have been going up.
>
> We have fires in California, in Florida, in other states, unprecedented fire season last year, directly correlated with higher temperatures, which dry out the soils, dry out the vegetation.
>
> We have a very serious threat of losing enough soil moisture in a hotter world that agriculture here in the United States would be greatly affected. . . .

The fact is that there is not one shred of evidence in the scientific literature, or in climate history, indicating that sea level could possibly rise more than three feet ("one meter") by 2017. The best estimate published by the United Nations Intergovernmental Panel on Climate Change (IPCC) in 2007, for the next 10 years, ranges between 0.8 and 1.7 *inches*.

The difference between Gore's conjecture about Greenland "in the next 10 years" and reality is stark. New satellites have found that Greenland is losing ice at a rate of 25 cubic miles per year. This information was published in *Science* in November 2006 by NASA scientist Scott Luthcke and many coauthors. The world's largest island has a total of 685,000 cubic miles of ice on it, meaning that the loss rate was measured at 0.4 percent per *century*. Gore had to know that. Any reference to Greenland's breaking up and slipping into the sea in 10 years is wild fantasy.

Despite this tiny increment of ice loss, these data, from a gravity-measuring satellite called GRACE, were greeted with some interest. It had long been thought that Greenland's ice was pretty much in balance, with the amount of ice accumulating in the center of its huge cap roughly equaling the amount being shed into the ocean. GRACE had indeed picked up an acceleration in the oceanic discharge.

But ice, like science, is pretty dynamic. A succeeding paper in *Science*, published by Ian Howat in early 2007, showed that the acceleration of ice loss detected by the satellite had reversed back to the presatellite rate, at least in the two major ice streams that Howat examined. Gore had to know that, too.

The IPCC'S 2007 "Fourth Assessment Report" on climate change includes a computer model projection for the loss of Greenland ice. It takes nearly 1,000 years to lose half its total. But the IPCC model assumes that the concentration of carbon dioxide in the atmosphere quadruples from its preindustrial background and then stays there for the entire time.

Right now, we're at 138 percent of the background, with an atmospheric concentration of 385 parts per million (ppm). Before the Industrial Revolution and for much of the period after the continents lost their massive Ice Age glaciers, the concentration hung around 280 ppm. It's highly debatable whether we could get to four times the background, or 1,120 ppm, even if we deliberately tried to do so. To maintain such a level for the next millennium assumes that we will still be burning fossil fuel—and at more than three times the current rate—in the year 3000. Even the Roman Curia wouldn't, in 1000 AD, have had the audacity to project the future state of the world for the next thousand years. Yet the United Nations (UN) blithely looks 1,000 years forward, making completely unfounded assumptions on energy use and human society. Many "thousand-year" political statements have been known to flop within a century, if not a decade.

Gore's statement about drought is wrong. He has to know that we have very good records about the area of the country under drought, back to 1895, reproduced here as Figure F. 1 (top). Figure F. 1 (bottom), on the same time scale, is the Northern Hemisphere's surface temperature history, from the IPCC. It's a waste of computing time to examine the correlation between the two in recent decades, because there isn't any.

Gore surely knew that, as the globe's temperature has risen since 1975, yields of almost all U.S. crops have increased significantly, and that they increased at a similar rate during the slight cooling of the Northern Hemisphere that took place from 1945 through 1975 (when some people worried about global cooling and a coming ice age). The American farmer is an adaptable creature, changing agricultural practices, and even crops themselves, faster than climate can change, and growing mainstay staples, such as corn, under a tremendous variety of climatic conditions. Many of our fresh vegetables come from a natural desert called California.

Ten years ago, Gore would have been called out for his remarks on *Larry King Live* as surely as he was for exaggerations about the Internet or embellishments about his college love life. But no more.

For many, the truth no longer matters when it comes to climate change. Science fiction movies such as *The Day After Tomorrow* or Gore's own *An Inconvenient Truth* cause the horde to clamor for action on global warming.

How did we get to such an extreme world?

We have written several books on this subject, contrasting facts, perceptions, and reality about global warming. Most recently, in late 2004, the Cato Institute published Michaels' *Meltdown: The Predictable Distortion of Global Warming by Scientists, Politicians, and the Media.* Since that writing, the political and physical climates have changed, as evinced by the preceding vignette. Michaels initially set out to simply revise *Meltdown*, but soon realized that so much new information has surfaced, and so many scientific changes have occurred, in

Figure F. 1

Percentage of the Lower 48 States Experiencing Severe (or Worse) Drought (top) and Northern Hemisphere Temperatures from the IPCC (bottom), January, 1900–August, 2008

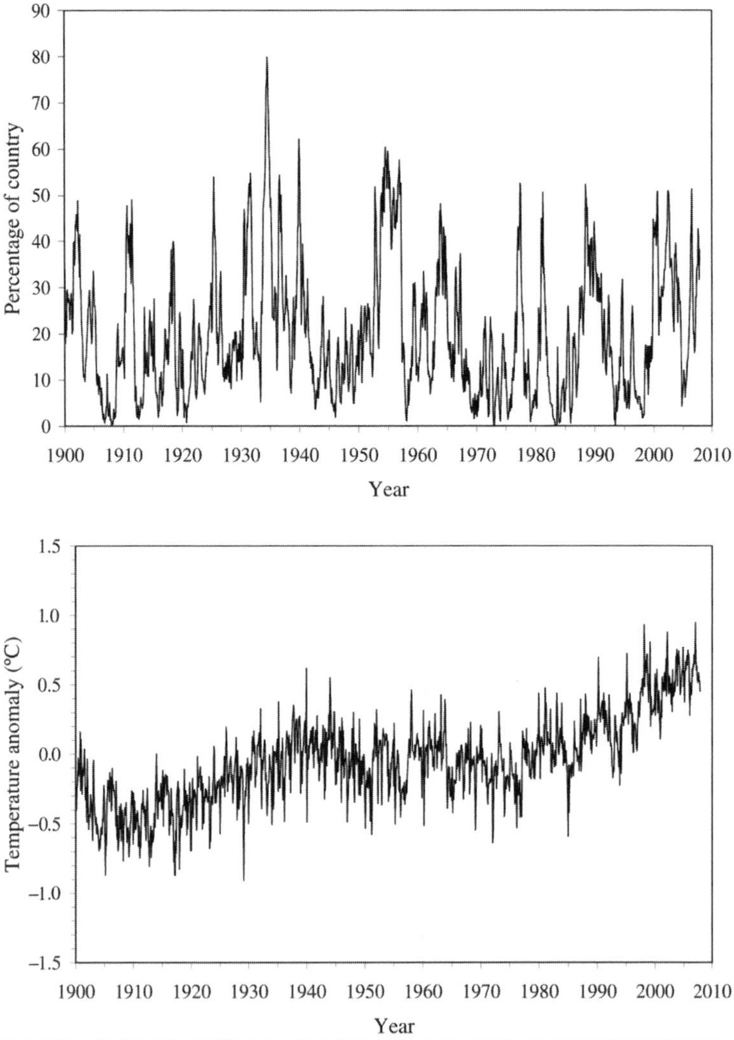

Source: National Climatic Data Center 2008 (top), http://www7.ncdc.noaa.gov/CDO/CDODivisionalSelect.jsp; Intergovernmental Panel on Climate Change 2007 (bottom) and updates.

a mere four years, that an entirely new book was required, one that would quantitatively analyze the scientific literature.

The rhetoric has changed. Discourse has degenerated into demagoguery. *Threatening* demagoguery.

Why has it become so politically risky to *not* view global warming as an unmitigated disaster?

In its larger incarnation, the political process is merely an instrument that adapts to public perception. Elected officials who do not echo popular perceptions risk losing the next election. Those who hold unpopular views on political matters, elected or not, no matter what the issue, will be ostracized because (1) they are an embarrassment to the process, and (2) they could conceivably change perception, forcing political flip-flops.

As a result, almost all that the public hears or reads about global warming is bad news. For the last two years, it seems that the Drudge Report has featured a global warming item almost every day. *Hurricanes are stronger and more frequent. Greenland is shedding ice at an alarming rate. So is Antarctica. Droughts and floods are increasing. If we don't do something drastic about our changing climate, it will soon be too late.* In 2007, actor Leonardo DiCaprio announced that global warming could ultimately cause the extinction of *Homo sapiens.*

Indeed, the political process has already responded. A 2005 energy bill, passed by a Republican congress and signed by a Republican president, mandated the production of massive amounts of ethanol from corn. It is easy to demonstrate, as did Tiffany Groode and John Haywood from MIT, that ethanol is a loser. It's used as a substitute for gasoline. But it turns out that the overall production cycle results in *more* carbon dioxide emissions than if one simply burned gasoline. Yet the ethanol mandate was sold as, among other things, a cure for global warming. That political process now has a clear mindset, and, needless to say, a lot of people are upset if they hear that ethanol won't solve anything and will actually contribute to global warming.

Then there is the charge that skeptical global warming scientists are "deniers" (named for Holocaust deniers), a peculiarly vicious label originally given to those who claim there's simply no such thing as human-induced global warming. We don't believe these people are correct, but we also haven't found one Nazi among them. They have their scientific reasons, although their argument is quite a stretch, given the nature of climate change in the last several decades.

How Perception of Extremes Evolves

How did the perceived climate of extremes develop? Is it because of a need, on the part of some scientists, to hype the lurid aspects of climate change at the expense of the more mundane? Do the publicity arms of universities or federal agencies, usually not stocked with scientists, get carried away with their rhetoric and emphasize extreme results?

Climate models, or compendia of models, usually give a range of expected temperature changes for doubling atmospheric carbon dioxide. But far too often, only the most extreme result enters the public discourse. Here's an example, courtesy of BBC Radio's Simon Cox and Richard Vadon.

It was January 2005, and Oxford University's David Stainforth and a large number of colleagues had just published a paper in *Nature*, which described a huge number of computerized simulations of global warming. Stainforth and colleagues created a virtual community of thousands of computer model users called climate*prediction*.net. They put out a a press release that only mentioned the most extreme value. Most of them predicted about 3°C (5.4°F) of global warming for doubling atmospheric carbon dioxide, but there were a very few outliers extending up to 11°C (19.8°F).

Climate*prediction*.net produced a press release about its work on January 26, 2005. There is only one sentence referring to future temperature: "The first results from climate*prediction*.net, a global experiment using computing time donated by the general public, show that average temperatures could eventually rise by up to 11°C."

"Up to 11°C." When the press dutifully reported this figure (and no other one, which was understandable, given that there was no other number in the press release), Myles Allen, an author of the paper and principal investigator for climate*prediction*.net, then blamed the press! "If journalists decide to embroider on a press release without referring to the paper which

(continued on next page)

(continued)

the press release is about, then that's really the journalists' problem. We can't as scientists guard against that." In fact, journalists did *not* embroider climate*prediction*.net's press release. They merely quoted it.

Cox and Vadon then presented several unnamed climate scientists along with the press release and the original paper. According to the BBC, "All were critical of the prominence given to the prediction that the world could heat up by 11°C."

One (unnamed) scientist told the BBC: "I agree the 11°C figure was unreasonably hyped. It's a difficult line for all scientists to tread, as we need something 'exciting' to have any chance of publishing . . . to justify our funding."

That doesn't happen every time, and plenty of scientists will be entirely straight when communicating to the public about the range of their climate results and their confidence in them. But in this case, climate*prediction*.net's press release did the world a disservice by using only one high figure. Ask yourself this: Which press release will get more attention, one saying that "most computer simulations of global warming predict a total warming of about 3°C (5.4°F)," or one that says, "The earth could warm by as much as 11°C (19.8°F)"?

Readers or viewers of news stories on climate change should beware every time they hear the phrase "as much as." There's obviously a range underneath the word "as," and, for some reason, the scientist, publicist, or reporter does not want you to know what it is.

Recently the definition has been expanded. The charge of "denier" is also thrown at those who argue that human-induced climate change is indeed real, but that this will not necessarily lead to an environmental apocalypse. And that's our stand. The data lead us to conclude that anthropogenic global warming (AGW) is indeed real, but relatively modest. We're not arguing against AGW, but rather against DAGW (*dangerous* anthropogenic global warming).

As Steven Hayward and Ken Green of the American Enterprise Institute have written, "Anyone who does not sign up 100 percent behind the catastrophic scenario is deemed a 'climate change denier.''' *Boston Globe* columnist Ellen Goodman wrote, in November 2006, "Let's just say that global warming deniers are now on a par with Holocaust deniers."

The existence of a mere hurricane now cries for a lynching. In December 2006, London *Guardian* columnist George Monbiot offered the view that "every time someone dies as a result of floods in Bangladesh, an airline executive should be dragged out of his office and drowned."

Even those who claim that there is little, if any, human influence on climate do not in fact deny the existence of climate change itself. The evidence for warmer recent times is incontrovertible. In the mid-19th century, glaciers threatened villagers in the Alps. Not 125 years later, the ice had retreated so far up the mountain that Julie Andrews, in *The Sound of Music*, crossed the Alps in dress shoes.

Perhaps all debate on climate change is irrelevant. After all, the standard argument from the political class in Washington is that "The science is settled," and that it's time to move on to policy. President Bush sees ethanol as a panacea for global warming, dependence on foreign oil, and international tension. So does Barack Obama. John McCain was the original author of S. 2191 (The name of John Warner [R-VA] was substituted when McCain became a viable Presidential candidate), a bill that mandated a reduction in carbon dioxide emissions to 70 percent below current levels by 2050. Sen. Barbara Boxer (D-CA), chair of the powerful Senate Environment and Public Works Committee, believes we can reduce emissions of carbon dioxide by 90 percent by 2050—only 42 years from now—if we simply pass a law saying we will do so. *How* we can accomplish this goal does not appear in any legislation or documentation, because no one in fact knows how to achieve such reductions.

How did we get to a world of apocalyptics and deniers, a world that is also one of impossible or ineffective policies on climate change? In other words, how did we get to such a climate of extremes?

The answer, it turns out, is purely logical. We bought it with our tax dollars, and we will pay the consequences for decades. There was no conspiracy (or at least no effective one). Rather, our extreme world of today is the result of "science as usual," hyped up on the steroids of massive public funding.

This book's first chapter describes the science of global warming, something that many readers already know by heart. If you do (and you'd rather not reminisce), skip to chapter 2, which describes our temperature histories and how they have changed over time. We detail six revisions to global records, each of which produces more global warming from the same original data. Having the revisions all in one direction, from three independent methods of temperature monitoring, is like tossing six heads or tails in a row. The odds are a little less than 1 in 50. So the continual upward-revising of warming trends in the same data possibly reflects something real that in reality is improbable.

Hurricanes are the subject of chapter 3, where we track the contentious controversy about whether or not they are made worse or more frequent because of global warming. Thanks to massive Hurricane Katrina's pillage of the Mississippi and Alabama gulf coasts (and her destruction of a criminally weak levy system in New Orleans), everyone seems to *know* that global warming was the cause, rather than merely being a passive bystander to the ruin of a city built several feet below sea level, literally sitting in wait of its destruction.

Chapter 4 deals with sea-level rise and melting ice, with particular emphasis on the disaster du jour, which is that Greenland is going to suddenly lose almost all its ice, perhaps before 2100, resulting in more than 20 feet of sea-level rise. It turns out there is hardly any data in support of this hypothesis, and an army of facts arrayed against it. But it is the specter of a Greenland disaster that is behind most of the current calls for dramatic cuts in carbon dioxide emissions. Antarctica also displays some screwy behavior that seems inconsistent and certainly confounds the myriad climate models that predict it should be warming smartly and experiencing increasingly heavy blizzards.

Forest fires, floods, and the various and sundry disasters associated with storms other than hurricanes are the subject of chapter 5. Here we detail some real whoppers laid on the public by elected officials you would think might know better. In the world of global warming, fact-checking has become fantasy, and perceptions have become the opposite of reality.

Chapter 6's title, "Climate of Death and the Death of Our Climate," refers to the phenomenon of warming-related deaths, such as the massive human die-off in France in the summer of 2003. It turns out that the weather anomaly that caused it was a tiny bubble of hot air embedded in a relatively cool summer around the planet. We describe the phenomenon of adaptation—something obvious to every economist but virtually ignored by every climatologist—in which succeeding heat waves kill fewer and fewer people, providing evidence that the response to changing climate is both political and technological.

In chapter 7, we describe ''publication bias,'' which is an attempt to answer the question, ''Why is all the news we read about global warming bad?'' There is a voluminous literature on natural biases in the scientific literature, and there are multiple causes. Curiously, climatologists claim to be immune from bias in their literature. But in making that claim, they are saying that they don't do something that virtually everyone else does, which is to publish more ''positive'' results (in this case, those fingering global warming for something) than ''negative'' ones (in which no relationship with global warming is discovered). There are several incentives for doing so, including the simple desire for (and professional requirement of) publication.

Chapter 8 proposes a modest solution to provide some balance between the mountain of bad news about climate change and the molehill of good news. We'll bet, for example, that if peer reviewers could no longer hide behind a cloak of anonymity, then a lot of biased shenanigans would stop. Finally, like any authors, we sum up our book with some stirring prose and bid you a good night's sleep.

1. A Global Warming Science Primer

Earth's mean surface temperature is doubtlessly warmer than it was 100 years ago. Get over it.

What matters is (1) how much it has warmed, (2) how much of that warming is caused by human activity, and (3) how the relationship between that activity and present temperatures can be translated into a reliable estimate of future warming and its effects.

The temperature changes. But so does the way in which temperature data are processed. We will demonstrate that fact in chapter 2. For now, however, we'll rely on existing histories.

Let's start out with a standard reference temperature history, the ground-based record from the United Nations' Intergovernmental Panel on Climate Change (IPCC) (Figure 1.1).

Figure 1.1
Global Temperature Departure from the 1961–90 Average, 1900–2007

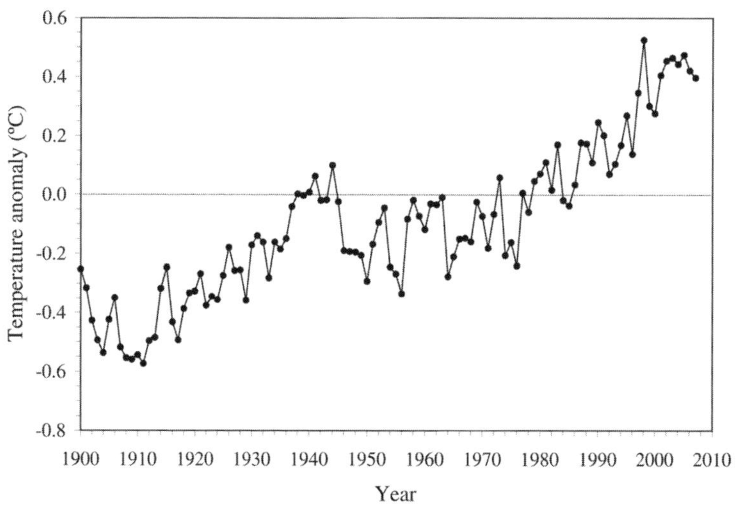

Source: IPCC 2007.

The IPCC history shows two distinct periods of warming, one roughly from 1910 through 1945, and then another that begins rather abruptly in about 1975. Their warming rates are statistically indistinguishable. In the last three decades ending in 2005, the warming rate was 0.178°C ± 0.021°C per decade (0.320°F ± 0.038°F). In the period 1916–45, the rate was 0.151°C ± 0.014°C per decade (0.272°F ± 0.025°F). Each of these is the observed trend plus or minus the statistical margin of error associated with it.

If those figures were the results of a political poll, the pundits would call it a tie—within the poll's range of error. Similarly, with temperature trends, adding in the "plus" to the first warming and subtracting the "minus" in the second reveals that the rate of warming in recent decades cannot be discriminated from the warming that occurred during a period of similar length in the early 20th century.

Their causes are very likely quite different, however. That said, one thing is for sure: the first warming was associated with a far smaller change in atmospheric carbon dioxide levels than the recent one. After all, we had not added very much carbon dioxide to the atmosphere before World War II.

Modeled vs. Observed Warming

Are recently observed climate changes consistent with computer models of climate change? That depends on where you look. If you examine surface temperatures observed at weather stations or as estimated from satellites, you'll conclude that the models can provide some quantitative guidance for the future. That doesn't mean that the models have all the answers, but it does suggest that they are largely sufficient.

There's another view—namely, that the models may have accurately captured much of the surface temperature change, but that they have missed the vertical dimension. If that's the case, then the match with surface temperatures is fortuitous—or worse.

Let's start with the first notion: that the models have something useful to tell us about future warming.

It's quite easy to demonstrate that the natures of the two periods of warming are quite different—and that the first one was probably caused by changes in the sun, whereas the second one has more of a relationship to human-caused emission of carbon dioxide and other greenhouse gases. We say "more of" because there are still other factors involved, such as a smaller solar effect and changes in land use, such as turning a "naturally" vegetated surface into a farmed one.

Greenhouse-effect warming occurs because certain constituents of our atmosphere, mainly carbon dioxide and water, are molecules whose shape allows them to absorb, and then release, radiation emanating from the earth's surface.

Bodies give off radiation that is proportional to their temperature. The hotter a body is, the more energetic the energy emitted. The sun, at 6,000°C (10,800°F), emits largely in the visible wavelengths of the universal electromagnetic spectrum (which is why our eyes evolved to "see" sunlight), as well as in the ultraviolet range (the energetic wavelengths that cause sunburn). The much cooler earth (with an average surface temperature of 15°C [27°F]) radiates largely in the less energetic infrared wavelengths (no one gets "earthburn"). Carbon dioxide and water vapor resonate with this low-frequency radiation and absorb some of it. The molecule reaches an unstable, physically "excited" state and then releases the packet of energy either up and out to space, or back down toward the surface. Consequently, greenhouse gases "recycle" the warming radiation of the earth in the lower atmosphere, resulting in a warmer surface and lower atmosphere than there would be in their absence. Another consequence is that the layer above most of the carbon dioxide—the stratosphere—cools because more radiation has been "trapped" below.

The mathematical relationship between the concentration of a greenhouse gas and surface temperature rise has been known for more than a century. The function is logarithmic, which means that the first increments of a greenhouse gas produce the greatest warming, and then increasingly large allotments are required to maintain that rate of warming. You can plot this function on your old graphing calculator or look it up on myriad websites.

Water vapor and carbon dioxide are known to behave quite similarly with regard to potential warming, so they can be (partly) considered to behave as the same greenhouse gas. As a result, atmospheres that are poor in both carbon dioxide and water vapor will respond strongly to the first new increments of either, because of the logarithmic nature of the temperature change. Again, increasingly large amounts of greenhouse gas would be necessary to maintain the same rate of warming.

Plenty of places on earth met this qualification before we put a lot of carbon dioxide in the air. Siberia and northwestern North America in winter are virtually devoid of water vapor: indeed, cold air can hold hardly any before it dumps it onto the ground in the form of frost or snow. It turns out that these are the places that have seen the biggest warming in recent decades. (Note: Antarctica, however, is not warming—a special case described in chapter 4). Further the warming rate in (dry) winter is much greater than it is in (moist) summer, consistent with greenhouse-effect theory.

Carbon dioxide concentrations in our atmosphere were approximately 280 parts per million (ppm) from the end of the last Ice Age to the beginning of the Industrial Revolution. Since then, they have risen to around 385 ppm, or a net increase of about 38 percent. In the 20th century, roughly three-fourths of the increase in atmospheric concentration took place after World War II.

There are several other emissions that alter the transmission of radiation through the atmosphere. On a molecule-for-molecule basis, methane, which is in much lower concentration, is 23 times more efficient at warming the lower atmosphere than is carbon dioxide. Its concentration has increased from about 875 parts per billion (ppb) around 1900, rising linearly to around 1,750 ppb by the 1980s. The increase was thought to have resulted from cow flatulence, coal mining, and leaky gas pipes (mainly in the former Soviet Union). Even so, none of these could possibly explain what happened after the late 1980s (see "Methane and the Perils of Scientific 'Consensus'").

Other industrial emissions are thought to counter the warming effect of greenhouse gases. A major cooler is something called sulfate aerosol—a particulate effluent emitted largely from coal-burning power plants. The relative cooling effect of sulfate is only "known" to a very broad range, from no cooling to nearly 2°C (3.6°F), which is very convenient, because it allows modelers to "choose" a value that, when added to the warming effects from carbon dioxide, methane, and a few other minor actors, forces a climate model's historical output to match the observed record shown in Figure 1.1.

At any rate, carbon dioxide still remains the biggest contributor to warming. A common counterargument is that most of the recent warming is a result of changes in the sun. But "solar" warmings should be a lot different from "greenhouse" ones. Rather than being

Methane and the Perils of Scientific "Consensus"

The increase in methane was remarkably constant from the early 20th century through the late 1980s. Every global climate projection assumed a similar increase would continue at least for another half-century.

No one disagreed. But, as if nature wanted to humble climate scientists, the rate of increase began to decline about 20 years ago, and the concentration of methane in the atmosphere has actually *dropped* in recent years (Figure 1.2). Nonetheless, IPCC's projections continue to show an increase (Figure 1.3).

Figure 1.2

Atmospheric Methane Concentration, 1983–2006 (top), and Change in Methane from Year to Year (bottom)

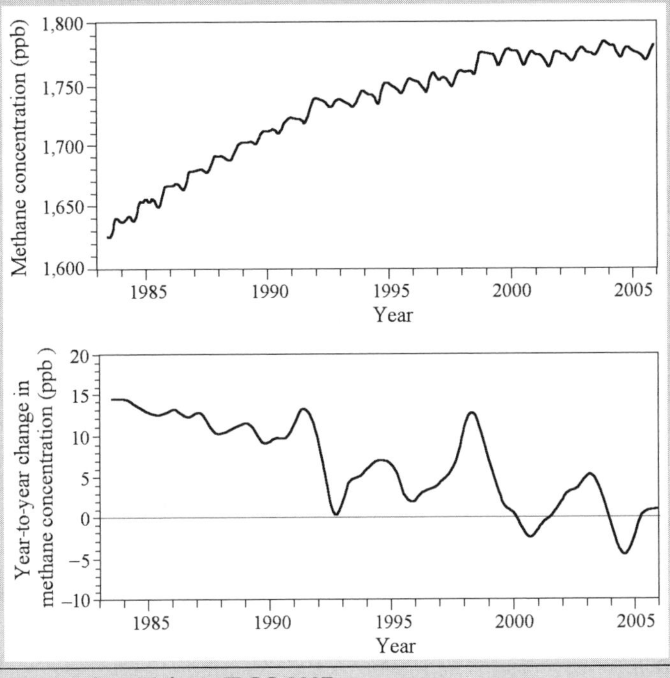

Source: Adapted from IPCC 2007.

(continued on next page)

(continued)

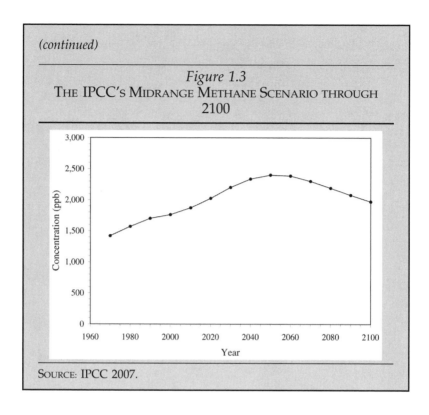

Figure 1.3
The IPCC's Midrange Methane Scenario through 2100

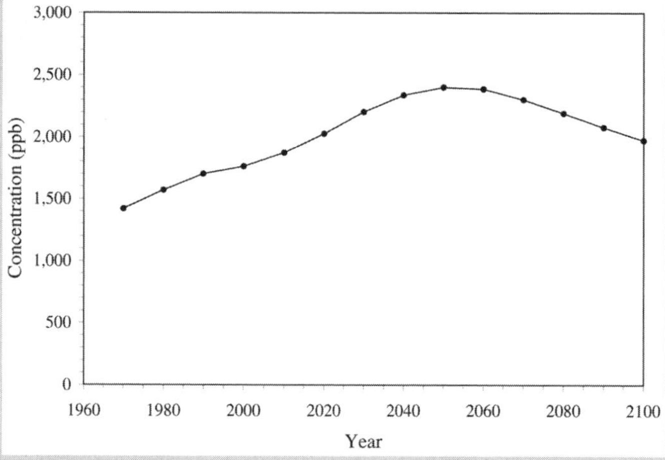

Source: IPCC 2007.

concentrated only in lower atmosphere, solar warming should be distributed in a way that is more uniform, heating both the lower atmosphere and the stratosphere, in which cooling has been observed in recent decades (see below for other complications!). Nor would a solar warming preferentially warm the winters so much as a greenhouse warming would.

An innovative analysis of U.S. temperatures illustrates the difference between the solar and carbon dioxide–induced warming.

As in the global temperature record (Figure 1.1), there are three distinct modes of behavior in the U.S. temperature history: a period of early-century warming, a midcentury cooling, and a final warming beginning in the 1970s.

The 365 black bars in each plot in Figure 1.4 are the rates of temperature change on the coldest night of the year (day 1) to the warmest (day 365). Note that these plots are *not* showing January 1

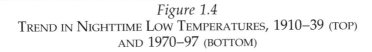

Figure 1.4
TREND IN NIGHTTIME LOW TEMPERATURES, 1910–39 (TOP) AND 1970–97 (BOTTOM)

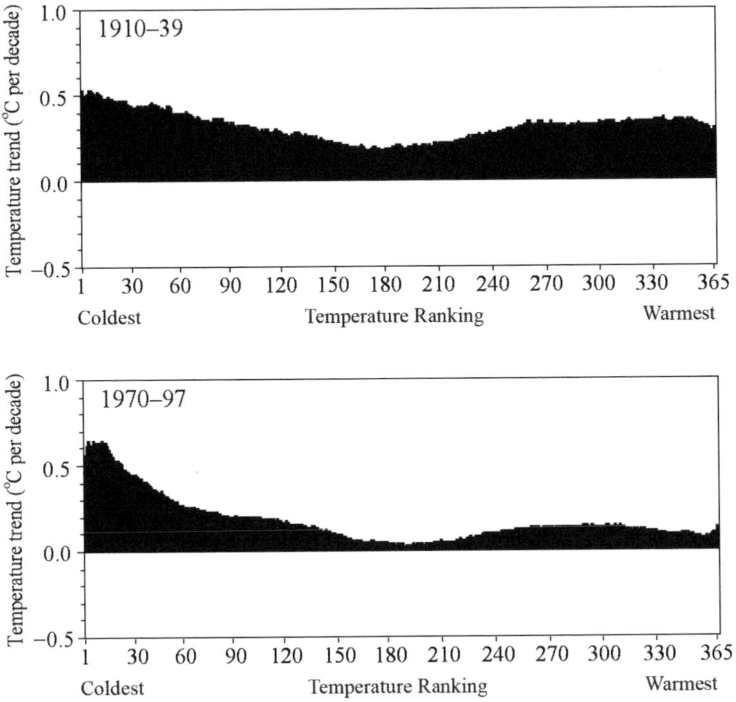

SOURCE: Knappenberger, Michaels, and Davis 2001.

NOTE: Day 1 is the coldest night, day 2 is the second-coldest, etc.

on the left through December 31 on the right; rather, they are arranging the data from the coldest day in each year to the warmest one. So the left side of each graph shows the trend in the coldest nights of the year, and the right side shows the trend in the hottest nights.

The top of Figure 1.4 is during the warming of the early 20th century (1910–39) and shows very little change in the trend of temperatures from the coldest (left) to the warmest nights (right). The bottom shows for the second warming, 1970 through 1997 (the last

Temperature Variability and Global Warming: Another Look

If the coldest nights are warming preferentially, then day-to-day variation in temperature must be dropping. Martin Beniston and Stéphane Goyette, of Switzerland's University of Geneva, recently published an investigation of the phenomenon of *decreasing* variability with greenhouse warming. They begin their article by noting, "It has been assumed in numerous investigations related to climatic change that a warmer climate may also be a more variable climate; such statements are often supported by climate model results."

They looked at low- and high-elevation temperature records for Switzerland and found the same thing we did for the United States—that the variability of temperature is decreasing, and that the decrease is concurrent with the increase in anthropogenic greenhouse gases.

They concluded:

> This investigation, carried out for a low (Basel) and a high (Saentis) elevation site in Switzerland, has shown that contrary to what is commonly hypothesized, climate variability does not necessarily increase as climate warms. Indeed, it has been shown that the variance of temperature has actually decreased in Switzerland since the 1960s and 1970s at a time when mean temperatures have risen considerably. Nevertheless, these findings are consistent with the temperature analysis carried out by Michaels et al. (1998), whose results also do not support the hypothesis that temperatures have become more variable as global temperatures have increased during the 20th century.

year in this particular study). Note how the coldest nights are warming up, much more than any others. This is the way greenhouse warming is supposed to work—and indeed is what has happened.

In the recent era, cold nights are warming much more so than hot ones. In other words, temperatures are becoming *less* variable. (A global examination of this phenomenon was published in our 2000 book, *The Satanic Gases.*)

We are not saying that the sun has had no influence on recent temperatures, but rather that the solar influence was clearly much greater during the warming of the early 20th century.

Nicola Scafetta and Bruce West, from Duke University, published an interesting paper along these lines in *Geophysical Research Letters* in 2006. Like many skeptical scientists, they prefer observed relationships to theoretical models. Scafetta and West examined the relationship between cycles in solar variations and cycles in temperatures using data back to the 17th century. Bottom line: "We estimate that the sun contributed as much as 45 percent to 50 percent of the 1900–2000 global warming and 25 percent to 30 percent of the 1980–2000 global warming."

Do the math. If 25 percent of recent warming is caused by the sun, and 50 percent of total warming since 1900 has the same cause, then 75 percent of the warming of the early 20th century should have had a solar origin. In 2007, using a different solar history and long-term temperature history, Scafetta and West duplicated their 2006 findings.

In sum, you can't throw the sun out completely when dealing with the recent warming, but it is not a majority contributor. That said, the bigger the solar impact, the smaller the human effect. The more "something else" is causing warming, the less sensitive the climate is to greenhouse emissions.

At any rate, the assumption that the majority of recent warming is from greenhouse changes remains the grounding rock of the notion that the models are providing some useful guidance with regard to 21st-century temperatures.

Because greenhouse gases tend to trap radiation close to the surface, there's less of a flux through the stratosphere, the layer of the atmosphere that begins about seven miles in altitude in our latitude. The stratosphere should cool slightly at the same time the surface warms. But if the sun gets warmer, so should the stratosphere. In fact, however, there is no record of stratospheric temperature that shows significant recent warming.

Both satellite and weather balloon data show stratospheric cooling, but carbon dioxide is only one cause. Changes in stratospheric composition owing to a slight loss of ozone have also contributed to cooling.

The ozone loss is hypothesized to have been caused by the breakdown of chlorofluorocarbon (CFC) refrigerants. The ban on these, a UN treaty known as the Montreal Protocol, is often cited as an example of successful global environmental regulation. If we managed to regulate CFCs, the reasoning goes, we can do the same for carbon dioxide. In reality the two are hardly analogous. CFCs are one of any number of chemicals that can be used for cooling, so substitutes exist; carbon dioxide, however, is the respiration of our fossil fuel–powered civilization. There is no politically and economically acceptable substitute currently available.

Nature of Observed and Future Warming

It is quite obvious from Figure 1.1 (and Figure 1.8, later in this chapter) that the rate of planetary warming since the mid-1970s has been quite constant (despite a lack of warming since 1998—the warmest year in the record). Computer models also tend to predict a constant rate of warming.

Figure 1.5 (see insert) is taken from the most recent IPCC report, published in 2007. It is the various warming projections from different computer models for the "midrange" scenario for future carbon dioxide emissions.

The IPCC's midrange scenario assumes that a "balance" of fossil and nonfossil sources of power evolves over the century, unlike its other scenarios, which are almost exclusively fossil-powered, or else presume the use of very little carbon-based energy by the end of the century.

Note that the projected rate of temperature change tends to remain the same once it is established (Figure 1.5; see insert); what the various computer models do is simply project *different* rates of constant change.

Figure 1.5 (see insert) also includes observed temperature changes from the IPCC's most recent iteration of the global history. (Note the discussion in chapter 2 about how this record itself has been altered and probably slightly overestimates recent warming). These figures are from the beginning of the recent warming, from 1977 through 2007. Note that they are also a straight line, but a line that tracks beneath the average of the climate models.

This would be the forecast of people who accept the fact that models tend to predict constant rates of warming (just different rates for different models), and combine that with the observed constant rate of surface warming, which yields a temperature change for the

The Wisdom of a Crowd (of Models)?

James Surowiecki's 2005 classic, *The Wisdom of Crowds*, demonstrates repeatedly that ensembles of independent estimates tend, over time, to perform better than a randomly selected individual one when asked to estimate some unknown quantity.

The common example pertains to the number of jelly beans in a jar. Whereas each of 100 people will almost certainly guess a different number, the real number will come close to the average of the 100 guesses.

People who teach weather forecasting are intimately familiar with this concept. Each student's forecasts are quantitatively evaluated, as are the average of all the students' forecasts. One student or another (or the group) may make the best forecast for an individual day. But when averaged over the long haul, the "group average" forecast is the likely winner.

This also applies to weather forecasting models. But it may apply to climate models only in a very special fashion.

In the daily weather forecast, it is well known that an "ensemble" of different computer models or different runs of the same model tends to perform better, over the long haul, than any individual model or run. That's because the models are indeed like individuals in a crowd in that they are "unbiased."

"Unbiased" means that the models (or the people) have no inherent problem that will make them systematically over- or underestimate temperature (or the number of jelly beans). But climate model ensembles can clearly have bias. For example, when climate models have been "intercompared" (as in the "Coupled Model Intercomparison Project" studies published initially by the U.S. National Center for Atmospheric Research's Gerald Meehl in 2000), they were all fed more carbon dioxide than was known to be accumulating in the atmosphere. Consequently, when compared with observed temperatures, they tended to predict more warming than was actually observed, a fact emphasized repeatedly in *Meltdown*.

(continued on next page)

(continued)

But both in those studies and in the model collection shown in Figure 1.5 (see insert), which uses a different carbon emission scenario (though the same one is applied to all models), the ensemble behavior resembles nature in that the warming is at a constant (rather than an increasing) rate.

This increases confidence in a forecast of warming that is indeed at the constant rate that has been observed for decades (subject to the post-1998 behavior described later).

21st century of about 1.7°C (3.1°F). This is about 40 percent less warming than the average projection given in Figure 1.5 (see insert).

There's a certain logic on behalf of the use of the models for some guidance for 21st-century temperatures, which can be summarized as follows: Both models and observations show a linear (constant) warming, but the observed warming is below the average model rate. Perhaps the "sensitivity" of temperature to changes in atmospheric carbon dioxide has simply been overestimated.

Has Global Warming Stopped?

Googling "global warming" will get you about 23 million hits. Most are of the gloom-and-doom variety. But not all of them. One thread that has emerged over the last year on many climate and policy blogs is that global warming "stopped" in 1998.

That's true (Figure 1.6) but caution is advised: 1998 saw one of the largest El Niños in recent history, and the associated suppression of the cold upwelling off of South America induced a huge temperature spike—one that was never exceeded in the subsequent decade. A plot of the year-to-year temperature change since then (Figure 1.6) clearly shows no obvious upward or downward trend.

That leads us to a fairly fearless forecast: The next big El Niño is likely to produce a temperature above that of 1998, resetting the global record. But in general, the same pokey warming trend that was established more than three decades ago will still be the rule.

In 2000, one of us (Michaels) published a paper in *Geophysical Research Letters*, showing that almost all the fluctuations around the warming trend that began in 1977 could be explained by the

"El Niño" in the Temperature History

Every global warming book will refer repeatedly to "El Niño," which has been blamed for floods, droughts, fires, diseases, and just about everything else—so many things, in fact, that Laurence Kalkstein, a well-known climatologist recently retired from the University of Delaware, used to deride it as the "Vitamin E" of climate.

El Niño is a slowing or even a reversal of the trade winds across the Pacific Ocean. No one knows exactly why it happens (the proof being that forecasts of impending El Niños are pretty lousy). Given that the trades are the largest single climate phenomenon on earth, slowing them has an awful lot of downstream effects, including spiking global temperature. It earned its name because there is a normal seasonal weakening of the trades that takes place in December—around Christmas—meriting the obvious title, The (male) Child. When an El Niño occurs, this normal weakening is extended throughout the year.

The trade winds are associated with a strong east-to-west current from South America, across much of the tropical Pacific. This current drags up cold water from beneath the surface, which is one reason why much of the Pacific shore of tropical South America isn't nearly as hot as one might think it should be.

When El Niño occurs, this cold "upwelling" is suppressed, and instead, the waters off of South America, and westward across much of the Pacific, become unusually hot. Needless to say, global average temperature rises. One of the biggest El Niños in the last 100 years occurred in 1998, and the temperature peak is quite evident in Figure 1.1. The year 1998 remains the warmest year in the global record, so warm that the succeeding decade shows no net warming trend whatsoever.

Of course, when El Niño stops and the cold upwelling returns, there's a lot of cold water waiting under the surface, and global temperatures drop. This phenomenon, not surprisingly, is called La Niña, and can be seen in the 1999 and 2000 temperatures.

(continued on next page)

(continued)

El Niño *is* actually correlated with a lot of weather anomalies pretty far from the tropical Pacific. For example, it usually (but not always) results in a much wetter-than-normal winter in Southern California.

Nature is pretty attuned to this natural fluctuation. For example, seeds of many plants in the Southwestern desert require the physical disturbance caused by a flood in order to germinate, so it's fair to say that El Niño makes the desert bloom. But (as described in chapter 5) the desert is, well, usually pretty dry, so that when El Niño goes away there's an unusually large amount of vegetation left to dehydrate and ultimately combust. So, though the chain of causation isn't rock solid (in some El Niño years, rainfall isn't enhanced), it seems plausible to blame an unusually vigorous fire-year in the Los Angeles basin on recent El Niño activity. Given that El Niños have been around forever (meaning many, many millions of years), Nature has been able to take advantage of their disturbance of normal weather regimes.

An El Niño year tends to be one in which global temperatures are elevated above a rather smooth trend. Will the next big El Niño year reset the global temperature record? And how important *is* a "warming trend" that takes over a decade to reset successive high temperature records?

magnitude of El Niño, changes in the sun's output as evinced by sunspots, and the amount of dust in the stratosphere contributed by big volcanoes.

Note the phrase "fluctuations around the warming trend." We're saying, whatever the cause (though it is probably carbon dioxide), there is a warming trend in the data, and the temperature changes around that trend are best explained by the other three variables.

So, to test if the warming trend has indeed "stopped," we ran our old model, which ended in 1997, and asked it to predict monthly temperature variations from either a continuation of the warming trend already established or a cessation of that trend at the end of 1997.

Figure 1.6
GLOBAL SURFACE AND SATELLITE TEMPERATURES, 1998–2007

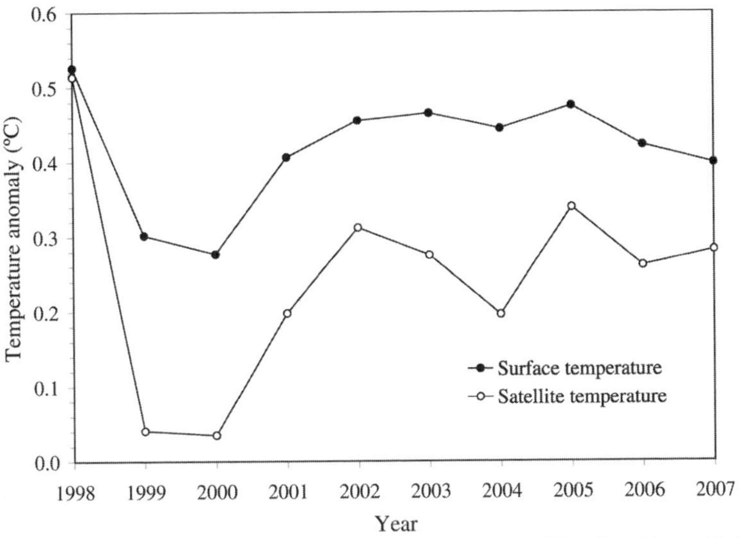

SOURCE: IPCC 2007 (surface temperature); University of Alabama-Huntsville 2007 (satellite temperatures), http://vortex.nsstc.uah.edu/data/msu/t21t/tltglhmam_5.2.

Which model predicted better? It was the one that assumed that the warming trend continued through 2008. Models that assumed temperatures were flat from 1998 to 2008 predicted surface temperatures to be lower than they were actually observed to be. In other words, El Niño and the sun conspired to halt the warming trend in the first decade of the 21st century. But in the future, they could behave in an equal and opposite fashion, as they did in 1998, creating a huge (but temporary) spike in global temperatures.

Rather than starting in the big El Niño year of 1998, perhaps it's fairer to start in 2001, after global temperatures recovered from the big El Niño–La Niña warming and cooling cycle. Figure 1.7 shows monthly temperature departures from average for two different records, the IPCC history and the University of Alabama-Huntsville satellite history (known as the UAH record). The two are offset

Figure 1.7

MONTHLY TEMPERATURE DEPARTURES FROM AVERAGE TEMPERATURE FOR THE IPCC RECORD AND THE UNIVERSITY OF ALABAMA-HUNTSVILLE SATELLITE, JANUARY 2001–JULY 2008

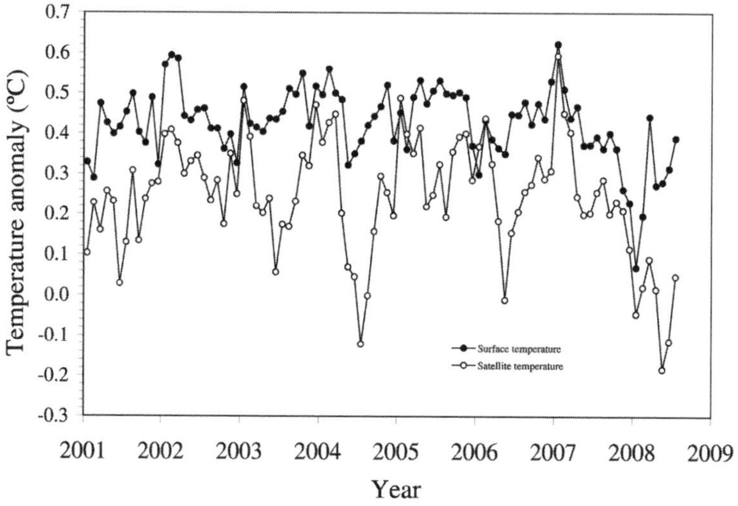

SOURCE: IPCC and updates 2007 (surface); University of Alabama-Huntsville 2008 (satellite), http://vortex.nsstc.uah.edu/data/msu/t2lt/tltglhmam_5.2.

because they are referenced to different averages. The IPCC is referenced to its 1961–90 mean, and the satellite record, which begins in 1979, is referenced to its 1979–97 average.

The period 2001–07 is the longest interval in which the IPCC record has shown no change since 1956–62. At the time, we had only increased atmospheric carbon dioxide 14 percent above its preindustrial value, compared with approximately 35 percent by 2006. Clearly the sun and El Niño are still capable of halting a warming trend, but they don't have nearly enough power to send temperatures back to where they were in about 1900.

Because of an additional finding, published in *Nature* in 2008 by Noah Keenlyside of Germany's Leipzig Institute of Marine Science, the implications of the recent lack of warming are remarkable. Keenlyside found that natural processes in the earth's oceans are likely to continue to offset much global warming through the middle of the next decade. If that is true, then we will have gone nearly two decades without any warming.

This is an arrow through the heart of the IPCC's "scientific consensus," and a serious blow to reliance upon the models. Take a look at Figure 1.5 (see insert). Is there a two-decade period in which *any* model predicts no warming? Obviously not! Aside from observed data, these models are our only guide to the future, and they clearly can no longer provide scientific cover for any policies predicated upon the notion of dangerous anthropogenic global warming (DAGW).

There's a further problem. The large warming that climate models produce is mainly a result of an increase in atmospheric water vapor that results from a much smaller warming produced by carbon dioxide itself. The source of that water vapor, of course, is the ocean. If the planet does not warm up for 20 years, there is a further, longer delay in the so-called water-vapor feedback, because the ocean cannot warm up instantaneously.

AGW (anthropogenic global warming), yes. But DAGW? We think not!

Reasons to Disbelieve the Models

The earth's atmosphere extends far above the planetary surface, and it is the vertical distribution of temperature—from the surface to the stratosphere (about 36,000 feet at our latitude)—that determines a lot of our weather. That zone is known as the troposphere, and it is where almost all the weather action takes place.

For example, when the difference between surface- and upper-tropospheric temperature is great, then the surface air is very buoyant compared with what is above it. Put simply, hot air rises and cold air sinks. The warmer the surface air, the more it is likely to rise. As a result, large amounts of air can bubble up. As air moves up, it cools, eventually to the point at which clouds form. The most common signature of a relatively warm surface overlain by a cold upper troposphere is the atmosphere's most visible bubble—the common thunderstorm.

Those who are skeptical of model projections point to a phenomenal mismatch between model predictions for temperatures above the surface and actual observations of them.

Is It Warming *Faster* than Predicted?

Much of the discussion in this chapter indicates that surface warming is taking place at a relatively constant rate (the current hiatus notwithstanding). But that's not what we read in one of the nation's most prominent newspapers.

A few years ago the *Washington Post*'s advertising slogan was, "If you don't get it, you don't get it." When it comes to global warming trends, it's the *Post* that doesn't "get it."

On January 29, 2006, *Post* global warming reporter Juliet Eilperin wrote that "[the] Earth is warming much faster than some researchers had predicted."

Where did this assertion come from? Certainly not from the earth's temperature history from the Intergovernmental Panel on Climate Change.

For the past 30-plus years—the period during which the earth's rising temperature has been most strongly associated with human activity—the average rate of warming (1977–2007) as measured by the IPCC record has been 0.168°C ± 0.017°C per decade (0.320°F ± 0.031°F) (Figure 1.8). Although there is a certain degree of annual variation around this trend, the overall rise has been incredibly steady; in other words, there is no appreciable trend to the trend (Figure 1.9). That means that the earth is warming at a constant, or linear, rate, *not* one that is accelerating. This is by and large the same behavior that the vast majority of climate models predict the earth's temperature will display when forced with ever-increasing amounts of carbon dioxide.

(continued on next page)

There's no doubt that getting the vertical temperature change right is central to accurately projecting the changes in weather that should accompany global warming. If the rate of temperature decline with height is projected to become smaller, then there will be fewer thunderstorms and a much more drought-prone world. If the opposite is true, the future is replete with lush vegetation fed by increasing

(continued)

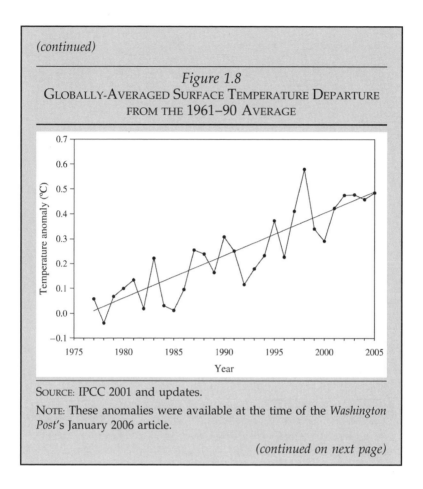

Figure 1.8
GLOBALLY-AVERAGED SURFACE TEMPERATURE DEPARTURE FROM THE 1961–90 AVERAGE

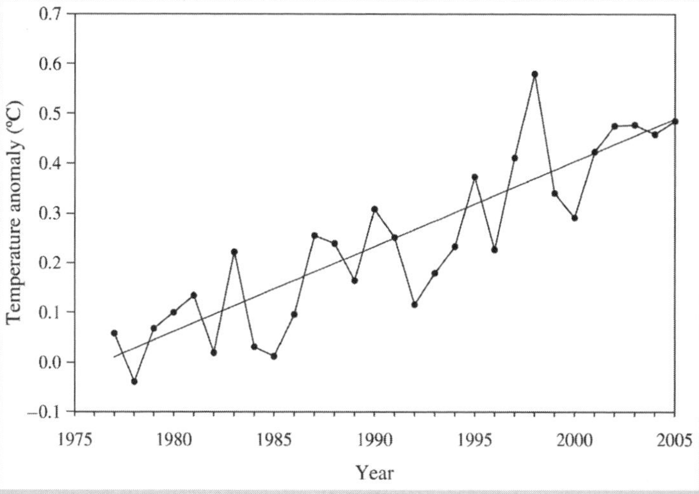

SOURCE: IPCC 2001 and updates.

NOTE: These anomalies were available at the time of the *Washington Post*'s January 2006 article.

(continued on next page)

rainfall during the growing season, when thunderstorms tend to occur.

The most recent (and very persuasive) evidence against the models was demonstrated late in 2007 in the *International Journal of Climatology* by University of Rochester's David Douglass and three colleagues—including John Christy, who developed the satellite-based temperature history discussed in chapter 2.

"The models are seen to disagree with the observations," Douglass et al. conclude. "We suggest, therefore, that projections of future climate based on these models be viewed with much caution."

(continued)

Figure 1.9
Year-to-Year Change in Annual Globally Averaged Temperature Anomalies, 1978–2005

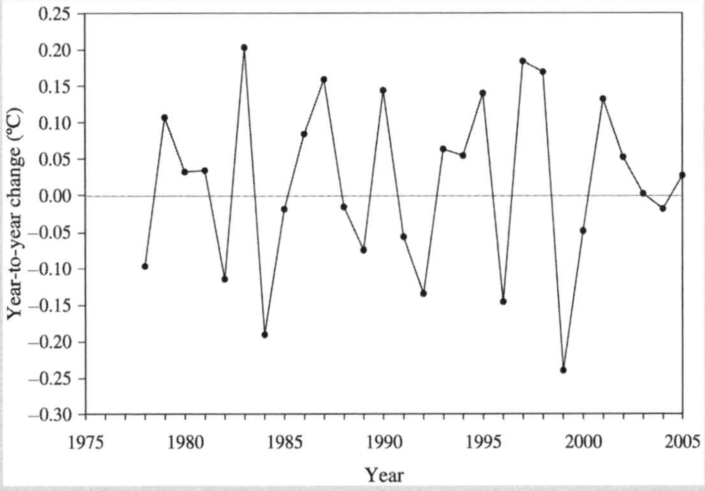

Source: IPCC 2001 and updates.

Who are Eilperin's researchers? If we turn to the "Third Assessment Report" of the Intergovernmental Panel on Climate Change (IPCC)—widely taken as the "consensus of scientists" at the time of Eilperin's article—it states, "The globally averaged surface temperature is projected to increase by 1.4°C to 5.8°C [2.5°F to 10.4°F] over the period 1990 to 2100." That is equivalent to a rise of 0.13°C to 0.53°C (0.235°F to 0.95°F) per decade. Compare that with the observed rate of warming we've established; clearly, the warming is running very close to the *lowest* end of the IPCC warming range.

Rather, the predicted mean warming rate is clearly higher than the *observed* one. Even NASA's James Hansen, the world's most quoted global warming scientist (and a person whom

(continued on next page)

(continued)

Eilperin has lionized in other *Post* articles), has argued that the warming rate over the next 50 years would be 0.15°C per decade ± 0.05°C (0.27°F ± 0.09°F), assuming only very modestly mandated changes in emissions.

Instead of hyping a nonissue, the *Post* would have done a far greater service by reporting in January 2006 that the earth's annual average temperature for the year 2005 fell *exactly* along the linear trend line established during the past 30 years (Figure 1.8) and as such, acted to further support the notion that the earth's temperature is warming up *less* than most people have predicted, assuming that the membership of the IPCC includes most climate people.

Climate models predict that the greatest warming should occur above the surface, not at or near the surface where we live. In 2000, the National Research Council examined this issue of the differential warming in various layers of the atmosphere and concluded that the surface was warming far more than the lower atmosphere; that pattern is not consistent with model predictions, and no obvious explanation was apparent.

The Douglass et al. team gathered output from models, surface observations, and balloon and satellite records, over the period 1979–2004, from which they calculated model-based and observed temperature trends at the surface and various altitudes in the tropical atmosphere. They focused on the tropics (20°N to 20°S) because "Much of the earth's global mean temperature variability originates in the tropics, which is also the place where the disparity between model results and observations is most apparent."

Trends from the models and observations agree at the surface but totally disagree from just above the surface to 14 kilometers (km) (8.7 miles) above the surface (Figure 1.10)

The models all predict far more warming around 10 km (6.2 miles) up in the atmosphere than is predicted at the surface. But all the observational evidence shows no such pattern whatsoever. In fact, there's a lot of *cooling* being observed at high altitudes rather than warming.

Figure 1.10

MODELED (FILLED CIRCLES) VS. OBSERVED (OPEN SYMBOLS) TEMPERATURE TRENDS FOR THE SATELLITE ERA (°C PER DECADE)

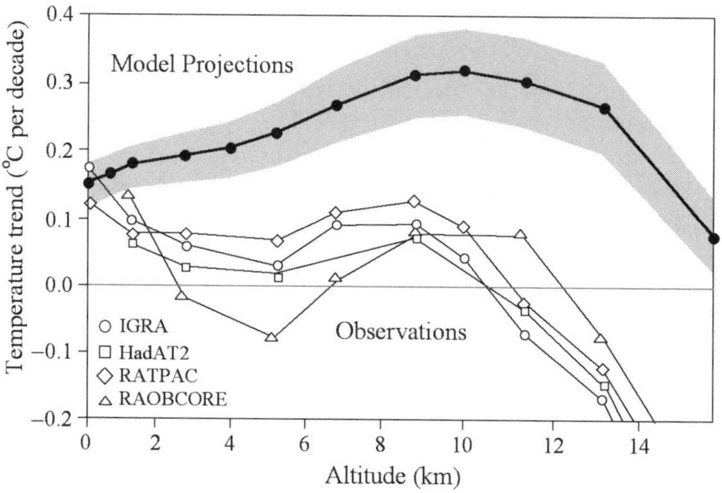

SOURCE: Adapted from Douglass et al. 2007.

NOTE: Observed temperatures begin in 1979. The model average comes from an ensemble of 22 model simulations from the most widely used models from throughout the world. The light gray area is the range of +2 and −2 standard errors round the mean from the 22 models, which is the 95 percent confidence band for the true model average. The acronyms refer to various observational databases.

Douglass et al. conclude:

> Model results and observed temperature trends are in disagreement in most of the tropical troposphere, being separated by more than twice the uncertainty of the model mean. In layers near 5 km [3.1 miles], the modeled trend is 100 percent to 300 percent higher than observed, and, above 8 km [5.0 miles], modeled and observed trends have opposite signs. On the whole, the evidence indicates that model trends in the troposphere are very likely inconsistent with observations that indicate that, since 1979, there is no significant long-term amplification factor relative to the surface.

The difference between surface and upper-tropospheric temperatures is increasing, not decreasing. The implications are huge.

For example, an atmosphere with a greater difference between the surface and upper layers is a more unstable one and will produce more precipitation.

An inaccurate precipitation forecast has huge implications for climate change predictions. Generally speaking, away from the high-latitude land areas (which are too cold to dry out much), places that get more rain have a wetter surface than those that do not. That means that more of the sun's energy is directed toward evaporation of water than toward a direct heating of the surface. (You can observe this phenomenon at the beach: Dry sand at noon will burn your feet, but wet sand will not).

So the amount of rainfall is a determinant of surface temperature. So is the amount of cloudiness. Everything else being equal, an atmosphere with more vertical motion (i.e., one where the surface is relatively warm compared with the upper layers) is one with more clouds. In the tropics, that means cooler days. Again, to specify the surface temperature correctly, it seems one has to get the vertical distribution of temperature correct also.

So how can the models get the surface temperature correct if they so dramatically miss the rest of the tropical atmosphere?

Intraday Temperature Issues

Clearly one signal consistent with greenhouse changes is an increase in the coldest temperatures, and that appears to have been observed (with the notable exception of Antarctica; see chapter 4). But the models have also overestimated vertical changes in temperature. Are there any other important aspects of climate change that they have gotten wrong?

One of the most prominent greenhouse-gas signals is the daily temperature range (DTR), which is the difference between the high and low temperature. Over most of the globe's land regions, that range has been declining over time—and the decline is thought to be a global warming indicator. Both maximum and minimum temperatures are rising, but the rise in daily low temperatures has occurred at a much greater rate, so the temperature range has gotten narrower.

This trend is related to increasing greenhouse gas levels because, everything else being equal, an atmosphere with higher greenhouse gas concentrations will have elevated nighttime temperatures. The

Figure 1.11
MODELED AND OBSERVED TRENDS IN MEAN, HIGH, LOW, AND DAILY TEMPERATURE RANGE, 1951–2000

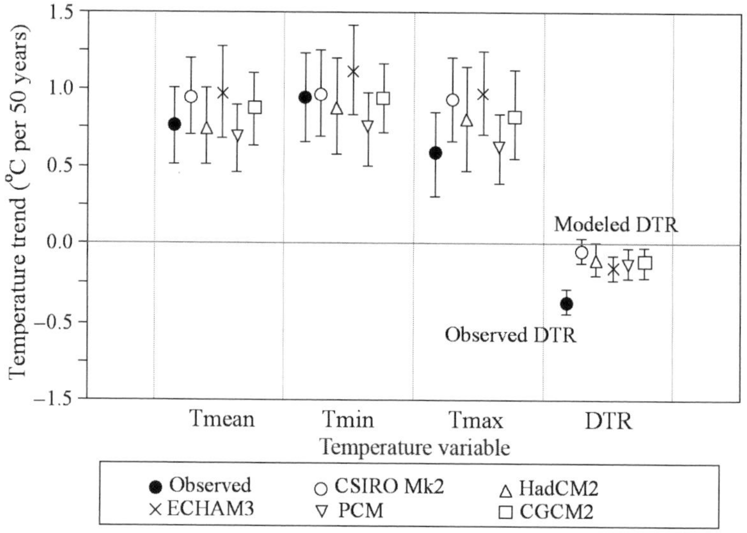

SOURCE: Braganza, Karoly, and Arblaster 2004.

NOTE: Acronyms refer to various models.

surface cools less at night because the earth's ability to radiate away heat from the lower layers is compromised by increasing greenhouse gases.

But climate models do not accurately replicate this effect. Take, for example, a 2004 study by Australian scientist Karl Braganza and two coauthors from the United States published in *Geophysical Research Letters*. The authors gathered data from all the global land areas with sufficiently long periods of record (forcing them to exclude Greenland, Antarctica, part of India, and most of Africa and South America), and compared the observed global decadal trends in maximum and minimum temperature and DTR with the output of five climate models in which the observed changes in 20th-century greenhouse gas and other atmospheric chemicals were simulated.

The results of the comparisons are summarized in Figure 1.11. Although the climate models, in aggregate, do a good job of reproducing the observed trend in minimum temperature, they overestimate the trend in maximum temperature. Each model does increase the daily high temperatures, but at a slower rate than the low temperatures. Actual observations show a much smaller increase in the daily highs. The net effect of that discrepancy on DTR is that none of the models can properly simulate the observed trend in DTR, which is declining at a rate greater than the models indicate it should be.

The critical issue here is that, given that DTR is really an indicator of greenhouse warming, the models must be mischaracterizing some very fundamental processes that are key to being able to accurately model our climate at all. In this case, the models can hardly distinguish between the rates of day vs. night warming, while, in reality, high temperatures are increasing more slowly than models predict them to.

The flaw in the greenhouse models may be related to cloudiness. Cloud cover over land areas increased during the last half of the 20th century. Cloudy afternoons are generally cooler than clear afternoons, so clouds could account for this large discrepancy between climate models and reality.

Of course, you could argue that you really can't model earth's climate without getting cloud cover correct, given that clouds have an awful lot to do with both planetary temperature and precipitation. You could even argue that, because of this cloud problem, the models might be getting the trends in minimum temperature correct by dumb luck, given that the fundamental physics are not correct.

The bottom line? Over global land areas, nighttime low temperatures are rising faster than daytime highs, and that trend is consistent with increasing greenhouse gas levels. Climate models are incapable of correctly reproducing the observed trends, and as a result are showing that daytime high temperatures are increasing faster than they are in reality. That error is present, in all likelihood, because the models have not properly captured some fundamental physical component of earth's climate.

Model "Tuning"

Can computer models be "tuned" to produce the right surface temperature? And could doing so make the upper layers in the computer model's atmosphere go haywire? Further, can aspects of a model be manipulated to give an expected output? How could that be done?

Go back to Figure 1.1, which is the IPCC's surface temperature history. Let's stipulate that it's correct (though the next chapter will raise plenty of questions). Carbon dioxide has been increasing throughout the 20th and 21st centuries, with relatively modest increases in the earlier years compared with what is being observed now.

If carbon dioxide were the sole driver of climate change, then the temperature would have changed in a similar fashion, with a constant rate of warming as carbon dioxide increases as a small exponent.

Obviously, the temperature history does not mimic what would be caused by the effect of carbon dioxide alone. That has been recognized for at least 20 years. *Sound and Fury*, Michaels' first book on global warming, cited a 1987 paper by Thomas Wigley that indicated that something other than carbon dioxide had to be influencing temperature. Whatever that "something" was, it had to enhance warming in the early 20th century, and then limit it or cause cooling in the midcentury.

That "something" is hypothesized to be finely divided particulate matter, usually in the form of sulfate aerosol. It is thought that such particles reflect away the sun's energy. The source: fossil fuels!

Fossil fuels, especially coal, contain some sulfur. When burned, the sulfur combines with oxygen, which, through a series of chemical reactions, ultimately appears as a finely divided dust, called sulfate aerosol, which is thought to create a cooling effect. Because there wasn't nearly so much coal combusted in the early 20th century as there is now, either carbon dioxide's or the sun's warming (the latter being more important than the former at that time) wouldn't be very attenuated by sulfates—not until the world industrialized, which was contemporaneous with World War II. And so, the story goes, sulfate cooling dominated carbon dioxide warming until the late 1970s, when carbon dioxide won the day.

This explanation is commonly invoked to explain the warming of the early 20th century, followed by a slight cooling to the mid-1970s, and the subsequent second warming which continued through 1998. Sulfur compounds emanating from coal-fired power plants are also thought to be responsible for (remember this one?) acid rain. So, the story goes on further, the sulfate effect was reduced (at least in North America and Europe) as "scrubbers" were put on the power plants to wash out the sulfur compounds before they could acidify precipitation. In other words, cleaning up coal enhances warming.

So there are two "knobs" on global warming models that can interact and produce something that mimics the surface temperature history.

One is the sensitivity of the temperature to changes in carbon dioxide, or the amount of temperature change expected for each increment of carbon dioxide. There's plenty of debate about exactly what this value is, so it can be specified as either high or low, depending upon the model. The other knob is the countering effect of sulfate aerosol. If the two knobs are adjusted just right, a model can show warming in the early 20th century, a cooling in the middle of the century driven by uncontrolled coal combustion, and another warming in the late 20th century as coal is cleaned up and carbon dioxide continues to increase.

The problem is that no one really knows the magnitude of the sulfate effects. Nor do we know precisely how the effects are distributed vertically. For example, sulfate aerosol is hygroscopic, meaning that it tends to gather water. Yes, that's right: It accomplishes "cloud seeding"—because water droplets cannot form unless they have a "condensation nucleus" to condense around. Simply put, sulfate aerosol should produce more water droplets in clouds.

The more cloud droplets there are, given a finite amount of moisture, the smaller each individual droplet is. And smaller droplets are more reflective, making whiter clouds, which should create even more cooling than would result from the sulfate itself. The brighter the cloud, the more the sun's energy is kept from reaching (and warming) the surface.

It therefore might be easy to specify the surface temperature by turning the carbon dioxide and sulfate knobs, though doing so might result in major errors in the vertical temperature calculation. How much of that has gone on is anyone's guess.

Those who seriously doubt the models have quite a point. "Believers" may be placing too much faith in the models because of (1) the apparent match with surface temperatures and (2) the fact that both observed and modeled surface temperature changes are occurring at a constant (rather than an increasing) rate. But the vertical temperature forecast errors make the match between the models and the surface history a possibly fortuitous result of model tuning.

The current state of global warming science is far from "settled." It's true that both modeled and observed *surface* temperatures are rising at a constant rate, but the models are clearly predicting too high a rate of increase. Again, perhaps the "sensitivity" of climate to carbon dioxide has simply been overestimated.

This is actually a minor problem, considering the problems with the vertical distribution of temperature and the daily temperature range. The former calls into question the scientific basis for any model projections of changes in cloudiness or rainfall. And if *those* are questionable, then any match with surface warming may be fortuitous. What's more, the fact that *none* of the IPCC's midrange models (Figure 1.5; see insert) generates a warming-free 15-year period in the 21st century, which is happening right now, is very disturbing.

Readers will note that we did not make a single argument for simply taking the model results at face value. That's because it is obvious that, in general, the models have predicted too much warming in recent decades.

Another noteworthy aspect of this chapter's discussion is that much of the work showing the problems with the models is "new" to our audience. Why has there been so little publicity about this good news? Do scientists—and the journalists who write about their work—tend to write about only "bad" news? Keep these questions in mind as you read the rest of this book.

2. Our Changing Climate History

It is obvious that the planetary surface temperature is higher than it was 100 years ago. But what about changes in the measurement and analysis of temperature itself? Have new ways of collecting information and analyzing it induced spurious warming or cooling into our weather histories? As the science of temperature sensing and the mathematical manipulation of those temperature data evolve, do the histories themselves change? And if they do, is the tendency to change in one direction?

Our model for the way science works would predict that over time, we will see more global warming in the same data. That's the "paradigm"-based view of science, first published in 1962 by Thomas Kuhn in *The Structure of Scientific Revolutions.* Kuhn proposed that most scientific research is conducted in service of existing "paradigms," or overarching philosophical structures that form a consensus view of science. The obvious one with regard to recent climate change is that carbon dioxide is the principal driver.

According to Kuhn, most scientific work tries either to explain anomalies in the paradigm or to show that anomalous data are in fact wrong. The use of sulfate aerosol to explain the obvious difference between a temperature record showing a relatively smooth increase in carbon dioxide and a temperature record showing warming . . . then cooling . . . then warming is a typical example of the Kuhnian view.

Is another one of Kuhn's dynamics in play regarding the temperature histories? Namely, that successive revisions will tend to get rid of more and more of that embarrassing midcentury cooling?

That's the subject of this chapter. Unfortunately, the devil is in the details!

There are three major ways in which the temperature of the surface or of the lower layers of the atmosphere is determined: from long thermometric histories at weather stations, from weather balloons launched simultaneously twice daily around the globe, and from orbiting satellites.

Surface Readings from the United States

We'll begin this discussion with surface thermometers. Specifically, we will start with the U.S. records, for several reasons.

The United States has maintained an extremely dense and high-quality network of thermometers back into the late 19th century. It is generally assumed that means that the U.S. temperature history is an accurate one.

In 2000, the National Research Council published a report discussing discrepancies between the surface, satellite, and weather balloon records. The panel found little disagreement between the U.S. surface temperatures the IPCC used (Figure 1.1) and those sensed by satellites over the United States. Over other parts of the globe, however, there were regions of substantial disagreement between surface and satellite data. Therefore, the U.S. surface temperature history is probably about as high-quality a record as there is.

The "Climatological Division" Record

The National Climatic Data Center (NCDC), a unit of the U.S. Department of Commerce, collects and maintains the U.S. climate history. There are many different networks and types of data. The longest record with the most detailed history comes from an aggregate of more than 16,000 stations (11,000 are currently active) that have been operated largely by volunteer "cooperative observers." (There are a few "professionally" monitored sites at airports and National Weather Service offices.) The "co-op" network was established in 1890; co-ops monitor temperature and/or precipitation, depending upon the station.

The tremendous advantage of this network is that it was specifically designed to monitor weather in a uniform fashion. Consequently, the type of instrumentation (thermometers or rain gauges) tends to be the same over time. There is, of course, evolution in technology, such as a switch that occurred from mercury-in-glass thermometers to electronic temperature sensors, known as the Maximium/Minimum Temperature System (MMTS). That changeover occurred mainly in the 1980s.

"Climatological Divisions" (CDs) are 344 multicounty aggregates in the lower 48 United States that are thought to have some geographic or climatic homogeneity. The CD data set is one of the least "massaged" of the U.S. records, and simply takes the large number of co-op stations within a CD and averages the daily high and low temperatures and 24-hour rainfalls.

There is one correction applied to the CD record, necessitated by "time-of-day" bias. Most—but not all—co-op observers record the previous 24-hour low temperature early in the morning, which is around the normal time that temperatures are at their low in the daily cycle. Imagine recording the temperature on a record-breaking cold winter morning. The result? Two record lows, one recorded at 7:00 am when a winter day's temperature is reset and another at 7:01 am. Afternoon observers tend to record after work, or after the time of the daily high temperature. Consequently, the likelihood of two consecutive record high readings is lower than that for two very low ones. And remember that there are more morning (cold) than afternoon (warm) observers.

The CD records are corrected to account for the percentage of morning observers, as well as for latitude and longitude of the CD (which determines how close to the morning low temperature a morning observation is likely to be).

Figure 2.1 gives the U.S. national average temperature based upon the CD data. Note that each CD is weighted for its relative size so that this represents a true national average (for the lower 48 states).

Each different analysis of national or global temperature data has its problems. In the CD record, the number and location of stations within each CD is not static. Nor is the environment surrounding each co-op station.

Note that the CD averages include co-op stations that are in cities or in which the environment may have changed from rural to suburban. CD averages also include National Weather Service stations, most of which are located at or near airports. Airports are usually built in rural locations. But a problem arises once they attract the inevitable related commerce (hotels, parking garages, etc.). Soon, airport areas resemble small cities, complete with the attendant "urban warming" that skews the temperature record.

Odd demographic factors can also bias temperature histories. Louisiana State University's Barry Keim closely examined CD data from New England and discovered that human migration patterns induced statistically significant changes in average latitude, longitude, and elevation of the stations within one Massachusetts CD. This particular CD extended from the western half of the Boston

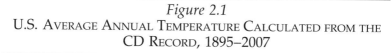

Figure 2.1
U.S. AVERAGE ANNUAL TEMPERATURE CALCULATED FROM THE CD RECORD, 1895–2007

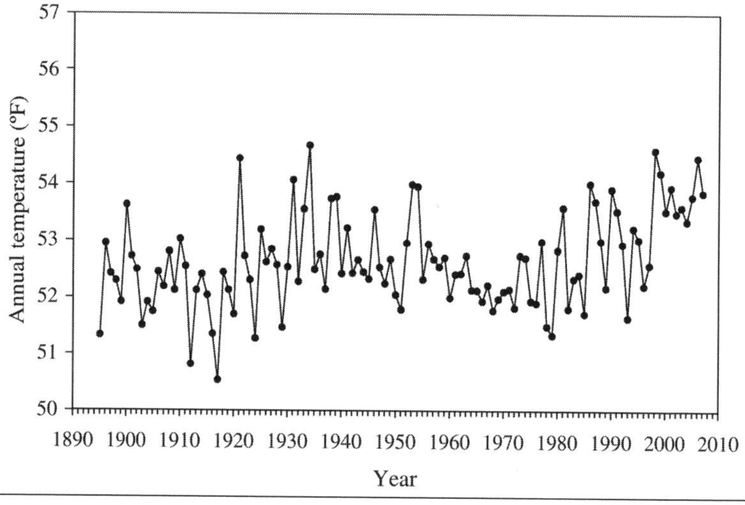

SOURCE: National Climatic Data Center 2008. http://www7.ncdc.noaa.gov/CDO/CDODivisionalSelect.jsp.

metropolitan area farther westward to the Connecticut River. Most of the CD is higher in elevation than western Boston, so the normal out-migration to further and further 'burbs raised the average elevation of the co-op stations. That would make that CD's readings *lower* than they should be. Other biases, including the general urbanization of the nation, would make them higher. Including city stations themselves in the CD record should mean that the record has some amount of artificial warming in it.

There are other factors that might give one pause before using the co-op data. We are a nation that used to cut trees down to clear land for agriculture. Consequently, the eastern United States was hugely deforested in the 19th century. But as agriculture (and people) moved West, the fertile soil and climate of the Midwest began to make eastern farming less and less competitive, so the East reverted back to forest. Many co-op stations that started off in the open might rather suddenly find themselves in the shade, as a nearby tree grows tall enough to cast an afternoon shadow. Obviously, that would induce an artificial cooling bias over time.

The U.S. Historical Climate Network

NCDC scientists, aware that there were multiple problems with the CD records, decided to create a history that they thought would contain fewer systematic errors. The U.S. Historical Climate Network (HCN) is a 1,221-station subset of co-op stations that are thought to be free from most common contaminations. They are generally in rural areas. An important feature is that each station was extensively examined for its history, including changes associated with station relocation. Records were examined to see if there was evidence of "discontinuities," such as might be caused by a newly extended shadow from a growing tree, general urbanization, or the building of new structures near the recording site.

Figure 2.2 shows the HCN and the CD data averaged over the United States. Both records begin in 1895. In the beginning of the record, the CD data tend to be warmer than the HCN, but by 1905 the HCN becomes warmer. The difference between the two over the entire length of record (Figure 2.3) is approximately 0.4°F (0.2°C). In other words, either the HCN is biased toward detecting more warming than there has been, or the CD record is somehow underestimating it.

Is there something inherently wrong with both records? Christopher Davey (Colorado State University) and Roger Pielke Sr. (University of Colorado) examined the 57 co-op stations in eastern Colorado. Ten of those are included in the HCN. They start by noting that the HCN does not specifically examine whether the exposure of the station corresponds to standards from the UN's World Meteorological Organization (WMO), which states that the site "should offer free exposure to both sunshine and wind [and not be] close to trees, buildings or other obstructions."

The *majority* of the HCN sites were in clear violation of the WMO standards. The Lamar station was sandwiched between two mobile homes, structures that tend to be lightly insulated and leak heat. Another one, in Eads, Colorado, was only 10 feet from another mobile home. One location was nothing short of ridiculous: "The Las Animas site had, by far, the poorest exposure for the USHCN that we visited," the researchers reported. The sensor was located six feet from a wall and an exhaust vent for a power plant!

Davey and Pielke couldn't conclude that the HCN was any better than the CD history. They found that the proportion of co-op stations with major siting problems was the same as in the HCN subset.

Urban Warming

It's long been noted that cities are warmer than the surrounding countryside. The bricks, buildings, and pavement heat up more than a vegetated surface, making daily high temperatures higher than those in the surrounding exurb. The surface—sidewalks and streets, highways and overpasses, skyscrapers, townhouses, parks, parking lots, parking garages, and so forth—is also much more uneven than that of flat farmland or gently rolling forest terrain, so that ventilating winds are less effective at dissipating the heat of the day.

There is no adjustment for urban warming in the CD history. The HCN data have been adjusted based upon either a population-based formula calculated by NCDC's Thomas Karl (HCN "Version 1") or an analysis of neighboring stations (HCN "Version 2"). If one station shows a warming trend that is not reflected in a neighboring one, the "warming" data are adjusted downward. Research published by one of us (Balling and Idso 1989) demonstrated that the "urban effect" (also known as the "urban bias" and the "urban heat island effect") was even evident at weather stations where the surrounding population was so small (2,500) that no one would have thought there *could* be an "urban" influence. Whether these changes are "caught" in HCN Version 2 is unknown, because NCDC has never explicitly published a list of precisely what corrections it has applied to individual stations.

The IPCC has used several different techniques to remove urban bias from its record. The original version looked for trends in neighboring stations. The current version simply adjusts temperatures downward by 0.0055°C (0.01°F) per decade, beginning in 1900.

People have tinkered with the CD history and removed some of the obvious urban stations, with only a minor (0.03°C to 0.06°C [0.05°F to 0.1°F]) reduction in net temperature change

(continued on next page)

(continued)

in the overall record. The IPCC claims that no more than 0.1°C (0.2°F) of their observed warming of 0.8°C (1.4°F) since 1900 is a result of urbanization.

The fact that HCN Version 2 warms more than the CD record (Figures 2.2 and 2.3), means that *other* corrections besides ones for urbanization that are applied to the co-op stations that make up the HCN are producing more "warming."

Figure 2.2

ANNUAL AVERAGE TEMPERATURE HISTORY FOR THE UNITED STATES, BASED ON THE CD DATA (OPEN CIRCLES) AND THE HCN VERSION 2 (FILLED CIRCLES), 1895–2007

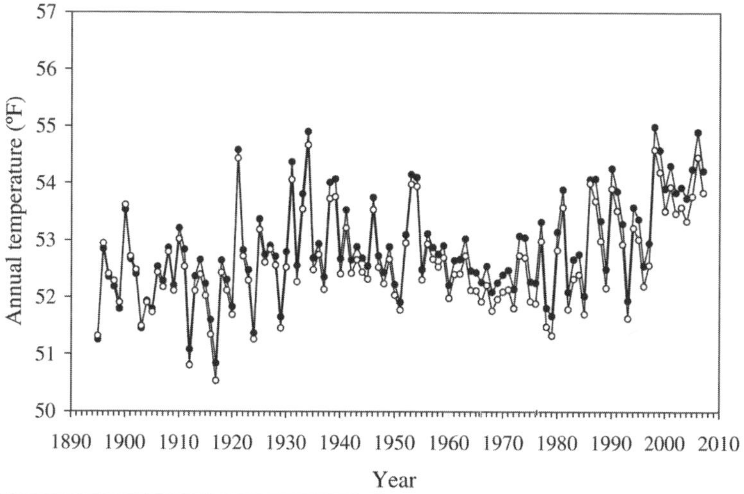

SOURCE: National Climatic Data Center 2008: http://www.ncdc.noaa.gov/oa/climate/research/ushcn/ (HNC); http://www7.ncdc.noaa.gov/CDO/CDODivisionalSelect.jsp (CD).

Most of the problems were because the sites either experienced poor ventilation (which would artificially raise both low and high temperatures) or were located over or near surfaces, such as blacktop or concrete, that would clearly affect daytime highs.

Figure 2.3
CD Temperature Subtracted from HCN Values

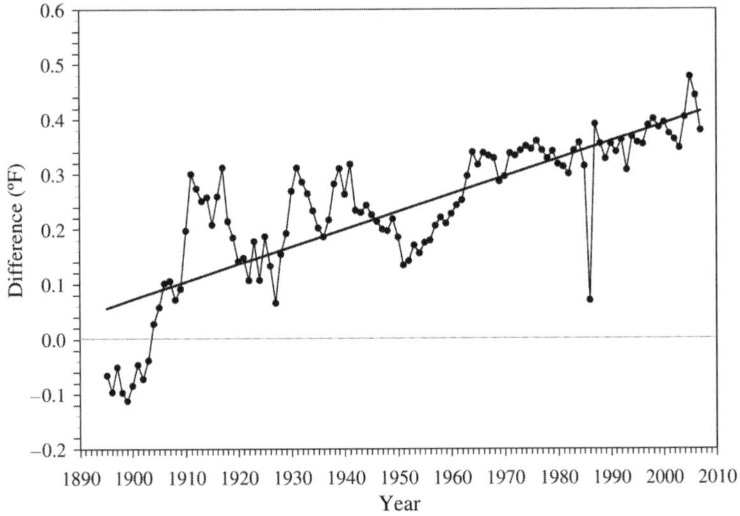

Source: National Climatic Data Center 2008: http://www.ncdc.noaa.gov/oa/climate/research/ushcn/ (HNC); http://www7.ncdc.noaa.gov/CDO/CDODivisionalSelect.jsp (CD)

Remembering that most of the co-op stations are rural, what accounts for the warm bias of the HCN compared to the CD record? A perplexing notion arises: Does the urban correction applied to the HCN data (which is not applied to the CD data) somehow *induce* that bias? Our sidebar "Changing Central Park's Climate Data" shows that circumstance could actually happen.

Global Histories

There are three global temperature histories from surface thermometers. The most cited is the record from the IPCC, also known as the CRU record because it originated from the Climate Research Unit at the University of East Anglia. The other two records are the Global Historical Climate Network from the U.S. National Climatic Data Center and the global history from NASA's Goddard Institute for Space Studies.

Changing Central Park's Climate Data

For our money, the best climate blog out there by far is Steve McIntyre's *Climate Audit* (http://www.climateaudit.org). McIntyre, a mathematician and former mining executive, uses the term "audit" because he believes there are a lot of Enron-like shenanigans going on in the climate community—a lot of fiddling with data and numbers and very little transparency.

As an example, he recently showed what all of the NCDC corrections have done to one of America's iconic weather stations: Central Park, New York.

Figure 2.4
CENTRAL PARK, NEW YORK, CO-OP STATION DATA

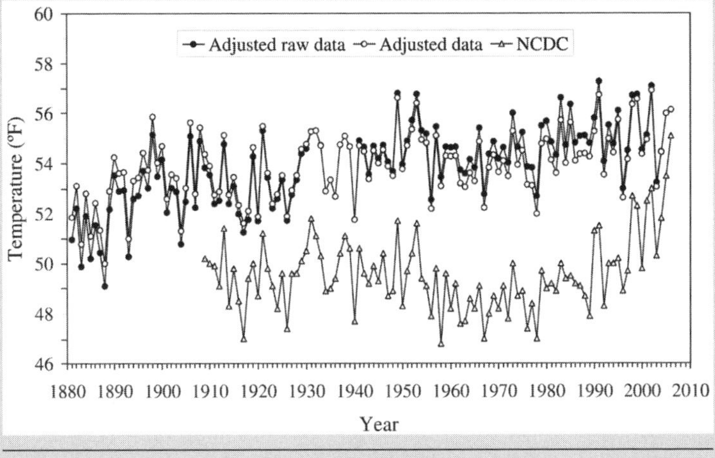

SOURCE: *Climate Audit* (http://www.climateaudit.org.)

NOTE: Closed circles are adjusted raw data for time-of-day bias and for station moves. Open circles (from NASA's Goddard Institute for Space Studies) are further adjusted to make the record compatible with surrounding rural stations. Open triangles are the NCDC-adjusted data based upon population.

(continued on next page)

(continued)

Go figure. Given that the raw data show no trend, there would have to be a massive assumed *decrease* in population to make the NCDC data suddenly warmer at the end of the record.

The Central Park history is an example of how much divergence there can be between temperatures actually measured on the ground and those that form our national and global histories.

All three are pretty similar in that they have two periods of warming, from 1910 to 1945 and then from 1975 to 1998, with an interval of slight cooling between the two. The most cited is the IPCC record.

This history was originally published by Phil Jones and several coworkers in 1985. Since then, it has gone through several iterations, which are displayed later in this chapter.

Unlike the HCN, the IPCC history does not correct for time-of-day bias, but it has some other very quirky corrections. The latest version is described in a 2006 paper in the *Journal of Geophysical Research* by Philip Brohan and several others, including Jones.

One adjustment is a "Homogenisation Adjustment," made when a station is moved. These adjustments occurred mainly in the 1940–60 period, when it was common for the official temperature-tracking site for a city to be moved from downtown to an airport. As a result, they *lower* the temperature of the pre-1960s data which, shown below, makes the mid-20th century temperatures colder. This has the effect of reducing the magnitude of the cooling between 1945 and 1975.

Anyone who lives around an airport knows that commerce soon migrates to that vicinity, with hotels, car-rental lots, and strip malls sprouting like mushrooms after rain. Consequently, the temperature should quickly rebound to the values measured in the previous urban location. The result is that, not only does this adjustment make the 1940–60 period cooler, it also probably makes the most recent years warmer.

Their urban adjustment is, to put it lightly, strange. It used to be that they considered rural–urban pairs, and when one (urban) station showed a warming trend that its (rural) near neighbor did not, then the urban station was either adjusted downward or removed. Some big cities, like Buenos Aires, showed no change and remained in the history.

They no longer do that. Instead, the temperature records are adjusted downward by 0.0055°C (0.01°F) per decade for the globe's entire land surface. That means there is the same urban bias assumed in both New York City *and* Antarctica.

As a consequence, our surface records are hardly "static." Figure 2.5 shows the difference between the last two iterations. It is very clear that the early years of the record have gotten colder. The result is more warming from the same data!

The observed rate of recent warming (1977–2007) is 0.167°C (0.301°F) per decade in the latest version.

Weather Balloon Records

Weather balloons are launched simultaneously around the planet, twice a day, to provide a detailed "snapshot" of the vertical structure of the atmosphere. They measure temperature, humidity, and altitude (barometric pressure). Wind is measured by tracking the flight from the ground. The data are then input into the giant computer models that forecast the weather up to 16 days in advance (no comment on how *good* that 16-day forecast is!).

The first temperature they record is at point of release—that is, the ground-based temperature. They then measure the temperature at various heights as they ascend. The instruments that record and transmit the data are called radiosondes.

Their purpose is send back accurate data. Consequently, the instrumentation is regularly calibrated. Even so, different nations use different sensing technologies, and even those that use the same instrumentation may process their information differently. Further, data transmission and sensing technologies evolve. Weather balloons are not designed to specifically determine our historical climatology, but they provide useful data.

James Angell, of the National Oceanic and Atmospheric Administration (NOAA), developed a temperature history from 63 balloon-launch sites worldwide, beginning in 1958. His history was first published in the journal *Monthly Weather Review* in 1975, and he continued to update it through 2005. The history includes

Figure 2.5
CURRENT ("HADCRUT3V") AND PREVIOUS ("HADCRUT2V") VERSIONS OF THE IPCC TEMPERATURE HISTORY (TOP), AND DIFFERENCE BETWEEN THE TWO (BOTTOM)

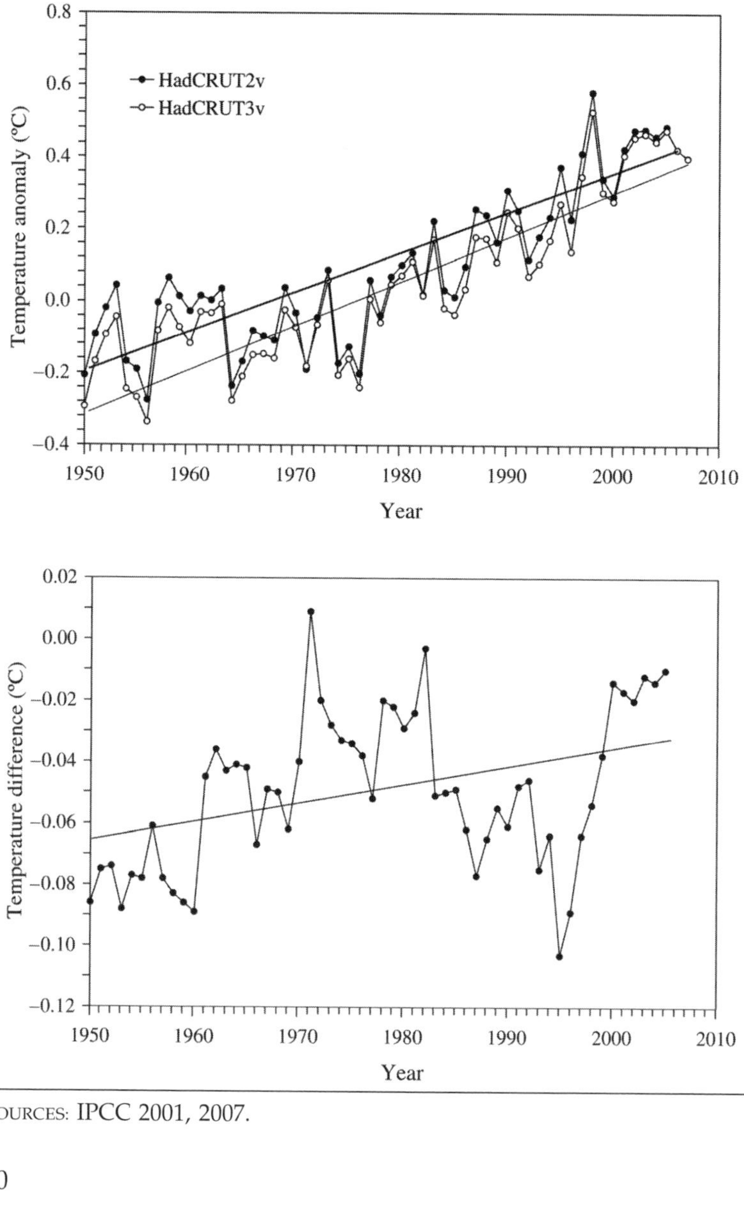

SOURCES: IPCC 2001, 2007.

the 5,000–25,000 foot layer, which should be free of any urban contamination.

As is the case for the surface temperatures, the data themselves have been revised. In 2003, Angell found that nine stations, all in the tropics, were unreliable because their temperatures varied far more than others from year to year. After noting this error and finding other problems, Angell published a new and expanded (85-station) history in 2005 that was carefully checked for changes in data quality and instrumentation. (The senior author was Angell's coworker Melissa Free, also of NOAA.) The new record was called RATPAC (Radiosonde Atmospheric Temperature Products for Assessing Climate).

The difference between the Angell record, which was the standard reference for decades, and RATPAC is considerable (Figure 2.6); RATPAC starts out colder and ends up warmer than the Angell record, adding a huge warming trend. The trend in the original data was 0.09°C (0.17°F) per decade, starting in 1958. The revised trend rose to 0.15°C (0.27°F), or 67 percent more warming than was in the original record.

Perhaps it's more important to look at the period from 1977 to the present, generally considered to be the era of greenhouse warming. In that case, the RATPAC warming trend is 0.160°C (0.288°F) per decade.

Satellite-Sensed Temperatures

In late 1978, NASA launched the first in a series of satellites designed to monitor global temperature from space. Those instruments are placed in orbits that measure the temperature at the same time of day globally.

The temperature sensors are called microwave sounding units (MSUs), and they actually measure the vibration of oxygen in the atmosphere, which is proportional to temperature. Different "channels" in the satellites can discriminate between different levels in the atmosphere.

The satellite record was first published by NASA scientist Roy Spencer and University of Alabama climatologist John Christy in *Science* in March 1990. They set off quite an uproar because the record showed absolutely no evidence for global warming.

Figure 2.6
RATPAC AND ANGELL GLOBAL TEMPERATURES FROM WEATHER BALLOONS (TOP), AND DIFFERENCE BETWEEN THE TWO RECORDS (BOTTOM)

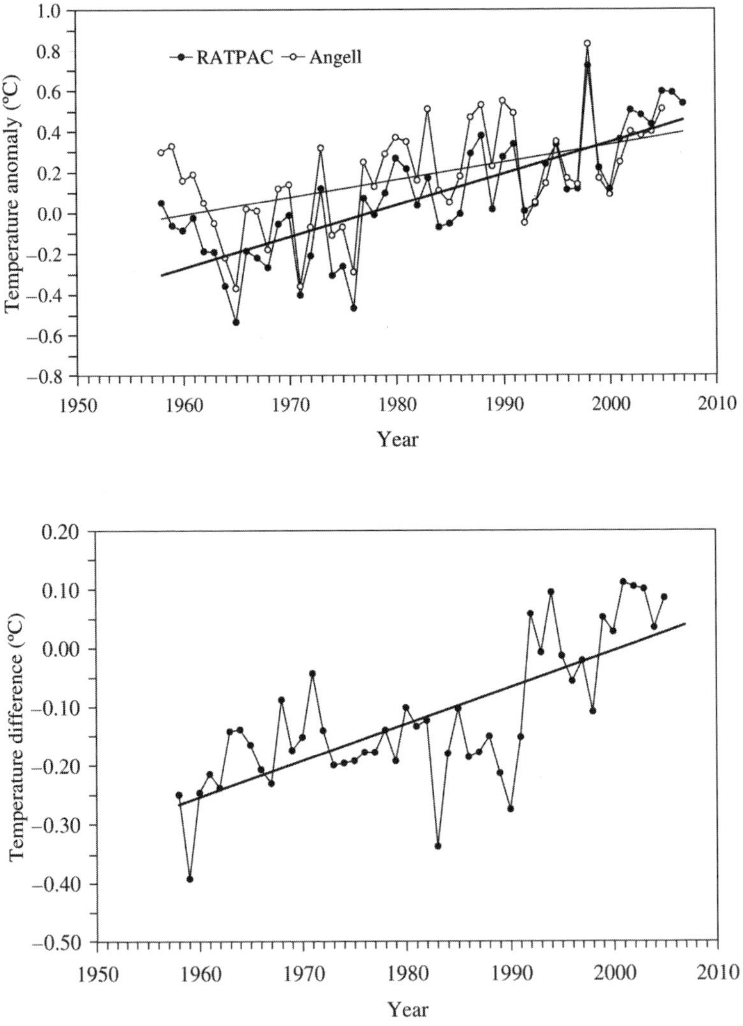

SOURCE: Angell and Korshover 1975 and updates, Free et al. 2005. Data are roughly the average from 5,000 to 25,000 feet.

The MSUs are functioning in the very harsh environment of space, and they are also subject to some pretty severe buffeting during launch and insertion into orbit. As a result, the individual sensors are employed only for a few years before they are replaced with a new satellite. As with weather balloons, calibration is critical. To make the records homogeneous, a new sensor must be calibrated against an existing one. That implicitly assumes that any "drift" in sensor sensitivity or response is known and accounted for, and that there are also no subtle changes in orbits over time that have not been detected and compensated for.

In 1997, Kevin Trenberth and James Hurrell of the U.S. National Center for Atmospheric Research, challenged the notion that the succeeding sensors had been properly calibrated against each other. Spencer and Christy took his objections into account and modified their history. But there was very little change—the satellite still showed no warming and, more important, was consistent with the weather balloon data measuring temperature in the atmosphere (roughly 5,000–25,000 feet).

In 1998, Frank Wentz and Mathias Schabel, from a small California consulting company called Remote Sensing Systems, published a paper in *Nature* in which they showed that the satellites' orbits were not so stable as had been assumed. Although the satellites were placed to sense temperatures at the same time around the planet, in fact, the orbits had been drifting.

Spencer and Christy began a log of the various corrections that were made because of orbital drifts, changes in the MSU sensors themselves, and other factors. Consequently, the MSU has become a highly "dynamic" data set, with slight changes applied once or twice a year. Spencer and Christy usually tweak the temperature trend by a hundredth of a degree (C) per decade or so.

Some of the corrections have been pretty large. In 2005 Carl Meaps and Wentz discovered errors in the way that Spencer and Christy were correcting for how the satellites varied on a 24-hour cycle. That correction, made in 2005, added a trend of 0.035°C (0.063°F) per decade.

Figure 2.7 shows three versions of the satellite data that were available at various times. Each correction can be applied to the entire data set, but if one record has been discovered to be too contaminated with errors, that record was be abandoned. As a result,

Figure 2.7
A COMPARISON OF THREE SATELLITE DATA SETS (TOP), AND TWO DIFFERENCES BETWEEN THE DATA SETS (BOTTOM)

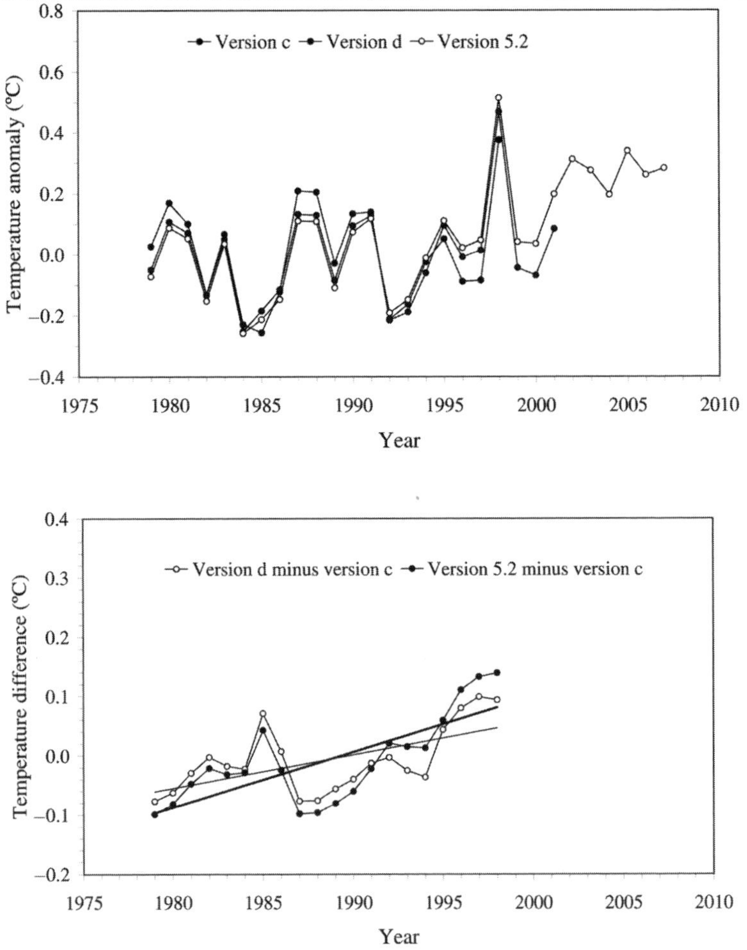

SOURCE: University of Alabama-Huntsville; Christy et al. 1998 (MSU 'c'); Christy et al. 2000 (MSU 'd'); http://vortex.nsstc.uah.edu/data/msu/t2lt/tltglhmam_5.2 (MSU '5.2'), 2007 (satellite temperatures), http:// vortex .nsstc.uah.edu/data/msu/t21t/tltglhmam_5.2.

version "c"—the original record, which is very close to the one published in *Science* in 1990—ends in 1998; version "d," corrected for orbital drift, ends in 2002; and version "5.2" is current and runs through 2007.

The changes that resulted are quite remarkable: Each major iteration of the satellite data produces more warming than the previous version. Figure 2.7 (bottom) is the difference between the intermediate ("d") record and the original ("c") record. There was no significant warming trend in the original data, but version "d" warmed at 0.075°C (0.135°F) per decade. Remember that these records can only be compared through 1998, when version "c" ends. Version 5.2 is very warm compared with version "c," with a trend of 0.095°C (0.171°F) per decade relative to "c." The *overall* trend in version 5.2 is 0.142°C (0.256°F) per decade.

A Strange Convergence

Think about it. Every "new" record we have examined here shows a greater warming trend than its previous iteration, over the same period of time. That has certainly had an effect on the public discussion and perception of global warming.

Let's start in 1995, with the three main records—the IPCC's, Angell's weather balloons, and Spencer and Christy's MSU satellite. (Figure 2.8)

You would notice several things in 1995. The IPCC surface temperatures appear to be quite constant through the mid-1970s, but since then shows a slight warming trend that appears to be around 0.2°C (0.36°F) per decade. The weather balloon data seem constant through the mid-1970s. They jump suddenly in 1976, but then *show little change from the late 1970s through 1995*. The satellite data, which begin in 1979, *show no warming trend whatsoever.*

You'll also note that there appears to be a great deal of agreement between the satellite and the balloon temperatures from year to year, except for a constant offset because they are referenced to different averaging periods. In other words, when one record goes up from year to year, so does the other, and by approximately the same amount. It is reassuring that two sets of data are in such fine agreement.

The IPCC surface history is the odd-record-out. There's some correlation with the balloon temperatures from year to year, but there is an apparent warming trend in the IPCC temperatures in the last 20 years that simply isn't reflected in the other two histories.

Figure 2.8
IPCC SURFACE TEMPERATURES (FILLED CIRCLES), BALLOON-MEASURED TEMPERATURES AT 5,000–25,000 FEET (OPEN SQUARES), AND SATELLITE-SENSED LOWER ATMOSPHERE TEMPERATURE RECORDS (WHITE CIRCLES), ACCORDING TO HISTORIES AVAILABLE IN 1996

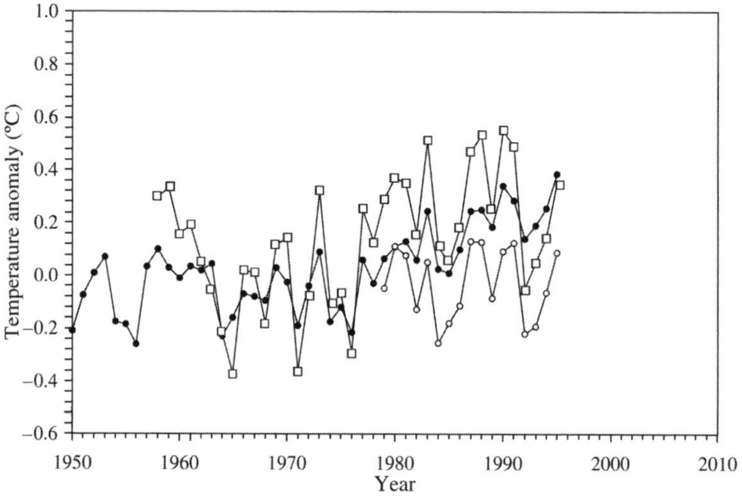

SOURCE: IPCC 1995, http://cdiac.ornl.gov/trends/temp/angell/angell.html (balloon); Christy et al. 2003 (satellite).
NOTE: Temperatures are expressed as departures from different mean values.

This discrepancy was often noted in public discussions of global warming. These were virtually all the data we had at hand: One record was in disagreement with two others, both of which were in agreement with each other.

Move forward now to 2000 (Figure 2.9). The early years of the CRU-IPCC surface records are actually a few hundredths of a degree colder than they were in the 1996 comparison. Consequently, there's a bit more "global warming" in the same history. The change resulted from the use of a different technique to transform the raw weather station data into a global average.

The difference between the IPCC and the other two records, if anything, has become greater. All three records clearly show the spike in global temperatures that occurred with giant 1998 El Niño,

Figure 2.9

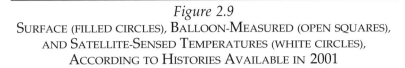

SURFACE (FILLED CIRCLES), BALLOON-MEASURED (OPEN SQUARES), AND SATELLITE-SENSED TEMPERATURES (WHITE CIRCLES), ACCORDING TO HISTORIES AVAILABLE IN 2001

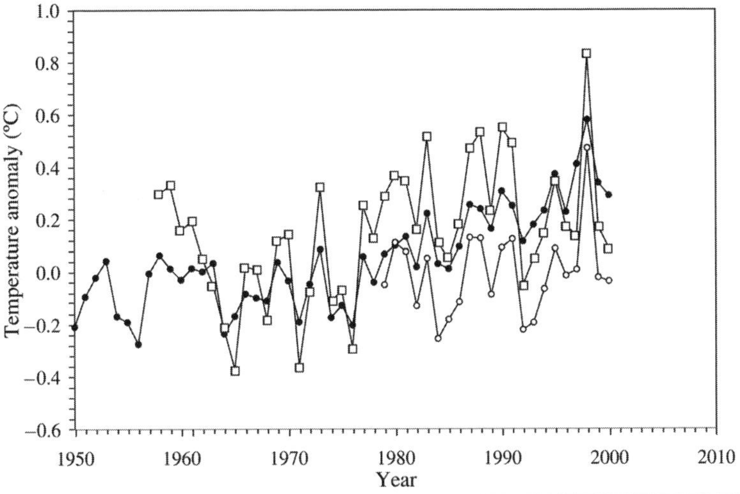

SOURCE: http://www.cru.uea.ac.uk/cru/data/tem2/ (CRU2v); http://cdiac.ornl.gov/trends/temp/angell/angell.html (balloon); Christy et al. 2003 (satellite).

but, after allowing for that one-year event, neither the satellite (since 1979) nor the balloon (since 1977) shows any warming trends that bear any resemblance to what is in the IPCC record.

Fast-forward to 2007. In the intervening period, Angell's RATPAC was published in 2005, and the Climate Research Unit at University of East Anglia (the source of the IPCC data) produced its 2005 revision. That one changed more than its last revision. We label it "CRUT3v," our shorthand for "third version of the CRU temperature record." (Figure 2.10).

The differences between the 1950–95 and 1950–2006 records are striking. The surface temperature record is even colder in the early years than it was in the previous iteration, and *both* the satellite and the balloon records now show warming!

Note that the early weather balloon data have become *much* colder in the early years. The result is more global warming.

Figure 2.10
SURFACE (FILLED CIRCLES), BALLOON-MEASURED (OPEN SQUARES), AND SATELLITE-SENSED TEMPERATURES (WHITE CIRCLES), ACCORDING TO HISTORIES AVAILABLE IN 2007

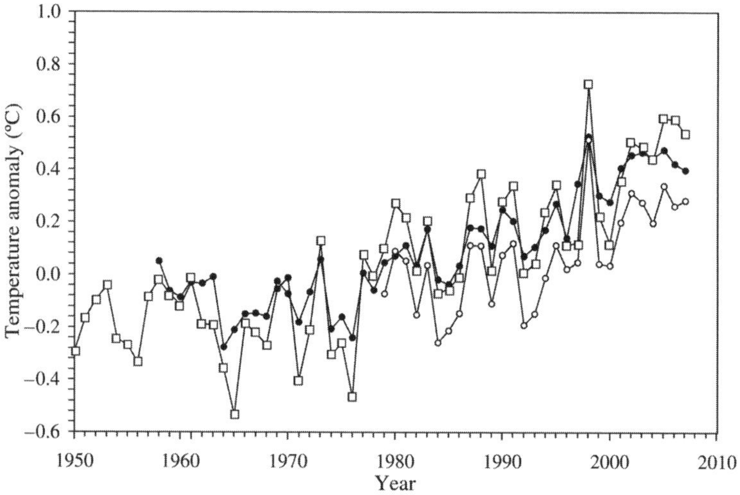

SOURCE: http://www.cru.uea.ac.uk/cru/data/temperature/ (surface); http://www.ncdc.noaa.gov/oa/climate/ratpac/index.php (balloon); http://vortex.nsstc.uah.edu/data/msu/t2lt/tltglhmam_5.2 (satellite).

Each of the three records are now in agreement with the surface, balloon, and satellite records showing warming rates per-decade of 0.167°C (0.301°F), 0.160°C (0.288°F), and 0.142°C (0.256°F) respectively, from the beginning of concurrency in 1979, values that are statistically indistinguishable from the other.

But Is It Real?

The IPCC considers its (changing) surface temperature history to be definitive and largely free from any systematic biases. As noted above, the temperature record is adjusted downward, beginning in 1900, for an arbitrary urban warming effect. The adjustment is linear and totals 0.06°C (0.11°F) by 2000.

Is urbanization all that could be contaminating the data and inflating observed warming? It has been known for years that landscape changes other than urbanization have an influence on temperature.

Figure 2.11
U.S. HISTORICAL CLIMATE NETWORK DATA THROUGH 2003 (SOLID CIRCLES), AND DATA ADJUSTED FOR THE WARMING BIAS FOUND BY KALNAY AND CAI (OPEN CIRCLES)

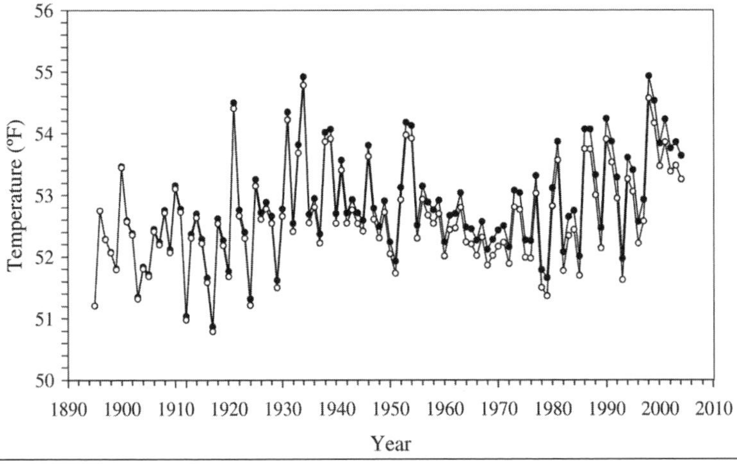

SOURCE: National Climatic Data Center, http://www.ncdc.noaa.gov/oa/climate/research/ushcn.html.

For example, the amount of solar energy that is absorbed and ultimately heats the lower atmosphere can change dramatically if a forest is transformed into a cornfield.

Eugenia Kalnay (University of Maryland) and M. Cai (Williams College) in 2003 performed a very interesting exercise in which they took advantage of the fact that weather balloon–measured temperatures show virtually no urban warming effect at a few thousand feet. Because we know how much that temperature, on average, changes with altitude, it is theoretically possible (and quite easy) to take temperatures measured aloft and "reduce" them to the surface value.

Figure 2.11 shows what happens to the U.S. Historical Climate Network (HCN) record when one does this. As noted above, the HCN has been revised; this figure uses "Version 1," the one that was valid at the time that Kalnay and Cai published.

The warming trend over the period of record (1895–2003) drops from 0.6°C (1.0°F) to 0.4°C (0.7°F), or one-third. If that were true on a global scale, and we attributed recent warming to changes in carbon dioxide, then that would have to mean that the "sensitivity" of temperature to carbon dioxide would be about two-thirds of what it was thought to be.

Climate scientists have been writing about problems with long-term temperature records for more than a century. Long ago, researchers noticed that temperatures in London were substantially higher than in the surrounding rural landscape; urban climatology has been a subdiscipline in the atmospheric sciences ever since. Recently, countless articles appeared in the literature on subjects as diverse as the urban heat island, changes in instrumentation, and changes in time of observation.

In 2004, one of us (Michaels) presented a paper at the Annual Meeting of the American Meteorological Society demonstrating that, although greenhouse warming is dominant in cold areas of the Northern Hemisphere in the winter, "economic" signals dominated elsewhere, especially in the summer. An expanded version of that paper was published in the journal *Climate Research,* in collaboration with Ross McKitrick from Canada's University of Guelph.

Other papers began to appear with similar findings. In 2004, Jos de Laat and Ahilleas Maurellis, both at the Earth Oriented Science Division of the National Institute for Space Research in the Netherlands, determined that local surface changes caused by industrialization accounted for a significant portion of global temperature increases in recent decades. They published their findings in *Geophysical Research Letters.*

De Laat and Maurellis used an idea similar to ours, in which they defined local carbon dioxide emissions as a proxy for the amount of local industrialization. They then divided the world into "industrialized" and "nonindustrialized" regions and calculated the temperature trends within each region. De Laat and Maurellis then repeated their analysis using a different cutoff value for what level of carbon dioxide emissions defined industrial and nonindustrial.

The results of their analysis are presented in Figure 2.12. They are quite striking, even if they aren't very surprising to others questioning the temperature history. Industrial regions with high carbon dioxide emissions have significantly larger warming trends than nonindustrialized regions, and larger trends than the globe as a whole. Similarly, as industrialization (as represented by carbon dioxide emissions) increases, so does the temperature trend. That is true

Figure 2.12

MEAN TEMPERATURE TRENDS (°C PER DECADE) FOR 1979–2001 FOR INDUSTRIALIZED REGIONS AND NONINDUSTRIALIZED REGIONS FOR DIFFERENT CARBON DIOXIDE EMISSIONS (TOP LINES); IPCC SURFACE TEMPERATURES (LEFT); SATELLITE-MEASURED TEMPERATURES, 0–10,000 FEET (RIGHT); AND CARBON DIOXIDE EMISSIONS LEVELS (BOTTOM LINES).

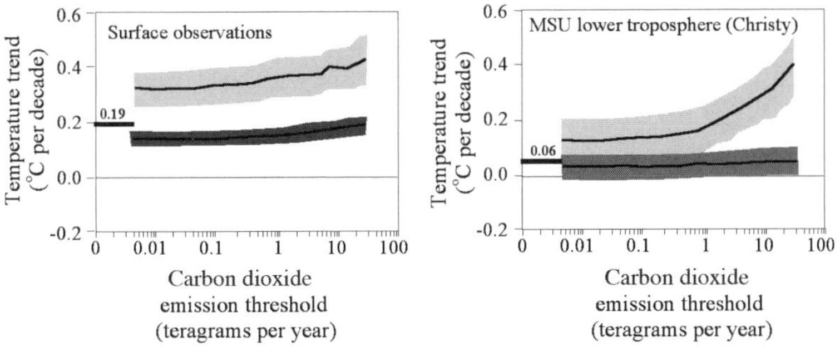

SOURCE: Adapted from de Laat and Maurellis 2004.

NOTE: The shaded regions indicate the uncertainties of the trend estimates. The thick solid bar inside the x-axis in each panel represents the global mean trend in each data set.

for both the surface and the balance of the lower atmosphere, or troposphere.

Note that this is not a measure of the local greenhouse effect—given that the concentration of atmospheric carbon dioxide is roughly the same around the world (there are some systematic geographic variations, but they are not large). Rather, it is more a measure of how much the local land surface has been altered.

They examined surface temperatures as well as two levels measured by satellite. Here we show only the surface and the 0-to-10,000-foot satellite record (labeled MSU, the sensor on the satellites), but there was a significant difference between the warming of industrialized regions compared with the more rural ones at all levels. Given that nonindustrialized regions show significantly smaller or even negligible temperature trends, the authors infer that a significant portion of the global warming temperature signal is localized—that is, confined to industrialized regions.

The authors also describe a serious flaw in the IPCC's surface temperature record. According to the paper, the "global" warming trend is about 0.2°C (0.36°F) per decade (it's actually 0.17°C [0.31°F] per decade for 1977 through 2007) but the data do not actually have global coverage. For instance, there's virtually no information from Antarctica, which is known to have cooled slightly in recent decades. When the authors calculate the satellite-based temperature trend for the same regions actually covered by the IPCC, they find that the IPCC's geographic selection results in an overestimation of warming by 33 percent. Applying this finding to the surface temperature data will reduce the "real" global warming to something around 0.12°C (0.22°F) per decade. It is interesting that this is the same reduction one gets by applying Kalnay and Cai's finding to the HCN.

If that finding is correct, a significant portion of surface temperature increase in recent decades has resulted from local surface-related processes as well as anthropogenic greenhouse gases.

Ross McKitrick and one of us (Michaels) published late in 2007 what we think is a comprehensive investigation into "nonclimatic" biases in temperature records; the article appeared in *Journal of Geophysical Research*.

We noted that more than 50 years ago, pioneering climatologist James Murray Mitchell warned that when using weather records to determine trends in climate: "The problem remains one of determining what part of a given temperature trend is climatically real and what part the result of observational difficulties and of artificial modification of the local environment." We maintained that

> two types of bias continue to affect the measurement of climate change. Observational difficulties, or data inhomogeneities (such as station moves and closure, record discontinuities, equipment change, and changes to the time of observation) are known to have affected records of mean temperature. Modification of the land surface, including urbanization and other economic activity, has been shown to affect local, regional and possibly global meteorology, and thus locally measured temperature data.

The IPCC assumes that there are many minor contaminants (besides urbanization, which it explicitly subtracts out) in the climate record, but that they are relatively inconsequential in the long run.

What a testable hypothesis just waiting for examination by skeptical scientists! It assumes that there should be no significant relationship between socioeconomic variables and trends in temperature over land areas. If significant relationships can be identified between socioeconomic variables and temperature trends, then other contaminants to the temperature records would be confirmed, and the IPCC's hypothesis must be rejected.

We examined the latitude–longitude gridded temperature data sets from the IPCC, and then we assigned to each grid cell information on gross domestic product, literacy, months with missing data, growth in human population, economic growth, and growth in coal consumption. We also added the satellite-based lower-tropospheric temperature trend to see if there was any local bias, sea-level pressure, a dryness index, proximity to an ocean, and latitude. These last four should filter out climate variability due to geographic factors. We looked only at land grid cells, because ocean temperatures should not be subject to economic or social biases.

The socioeconomic signals in the temperature trend data were loud and clear.

We concluded that our results were

> consistent with previous findings showing that nonclimatic factors, such as those related to land use change and variations in data quality, likely add up to a net warming bias in climate data, suggesting an overstatement of the rate of global warming over land.

We found that the data were pretty good over much of North America with the exception of northern Canada and Mexico (two economically poor regions). Where poverty was pervasive, over Africa and much of south Asia, there was clearly much less warming than temperature records indicated. Figure 2.13 (see insert) shows our findings.

Figure 2.14 shows the frequency of observation of different rates of warming in the IPCC surface and the satellite data, and in IPCC surface data adjusted for the biases we found. A value of, say, 0.1 to 0.2 means that the observed trend was between 0.1°C and 0.2°C (0.2°F and 0.4°F) per decade. The y-axis is the relative frequency (number) of trends of various magnitudes. Interestingly, our "adjusted" data look a lot more like the satellite data than the IPCC's original record. Inspection of Figure 2.14 reveals that the biggest

Figure 2.14
DISTRIBUTIONS OF TEMPERATURE TRENDS, 1979–2002: SURFACE (IPCC) (TOP); TROPOSPHERE (SATELLITE) (MIDDLE); AND ADJUSTED SURFACE (BOTTOM)

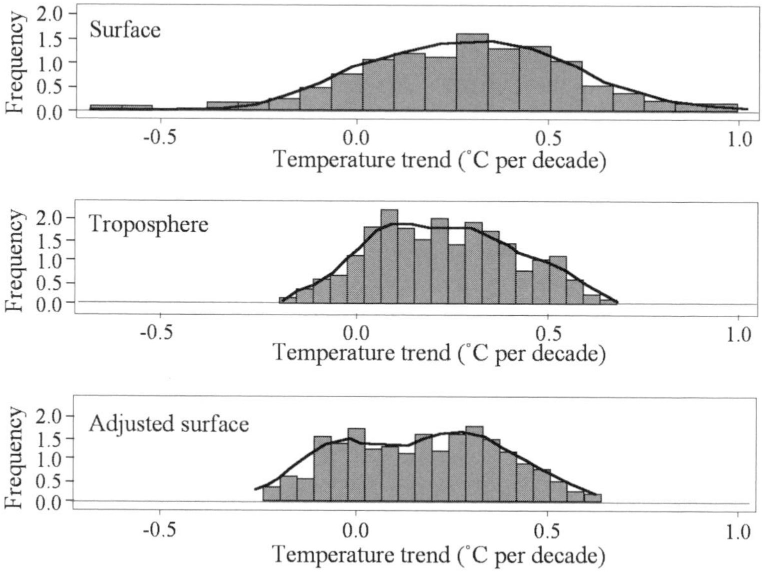

SOURCE: McKitrick and Michaels 2007.

change was that our adjustment lopped off the very warm right-hand "tail" of the IPCC's temperature distribution. In other words, the places showing the greatest warming had the greatest nonclimatic bias in their records.

Our conclusion:

> Nonclimatic effects are present in the gridded temperature data used by the IPCC and they likely add up to a net warming bias at the global level that may explain as much as half the observed land-based warming trend.

Remember that this does not mean that half the entire planet's warming may be spurious. One has to balance the fact that the land surfaces have warmed more than the ocean, but that the land is only about one-third of the total surface. As a result, the global warming rate since the late 1970s drops from the familiar 0.17°C (0.31°F) to

0.13°C (0.23°F) per decade, pretty much the same number you get if you apply Kalnay and Cai's findings to the HCN, or the calculation of de Laat and Maurellis.

After our findings were published, not one news article appeared noting that the amount of "real" global warming since the late 1970s is nearly 25 percent less than previously thought, and that this was consistent with two other independent studies. Imagine what would have happened if we had found an extra 25 percent of warming!

"Your Aim Is to Find Something Wrong with It"

Perhaps the most fitting vignette demonstrating the "Climate of Extremes" in global warming science comes from the main author of what is now the IPCC's temperature history.

Scientists often call the IPCC history the "Jones and Wigley" record, because of several landmark papers describing it published by Phil Jones of the University of East Anglia and Tom Wigley, now at the U.S. National Center for Atmospheric Research in Boulder, Colorado.

In its "Third Assessment Report" in 2001, the IPCC gave the 100-year surface temperature trend as 0.6°C ± 0.2°C (1.1°F ± 0.4°F). Australian researcher Warwick Hughes became interested in how this error was calculated. Was it because of errors inherent in the raw data? How were confounding effects, such as the growth of cities, accounted for? How about changes around the sensing equipment, such as the erection of a building?

So, Hughes wrote to Phil Jones, asking for the original temperature data. On February 21, 2005, Jones responded:

> We have 25 years or so invested in the work. Why should I make the data available to you, when your aim is to try and find something wrong with it?

Normally science thrives on the free exchange of data. But not in a climate of extremes.

3. Hurricane Warning!

> "Global warming isn't to blame for the recent jump in the number of hurricanes in the Atlantic, concludes a study by a prominent federal scientist whose position has shifted on the subject."
>
> —Associated Press, May 19, 2008

> "After some prolonged deliberation, I have decided to withdraw from participating in the Fourth Assessment Report of the Intergovernmental Panel on Climate Change (IPCC). I am withdrawing because I have come to view the part of the IPCC to which my expertise applies as having become politicized."
>
> —Hurricane scientist Christopher Landsea,
> in an "Open Letter" to his scientific colleagues,
> January 17, 2005

Hurricanes and global warming are a hot item, but the relationship became controversial even before Katrina demolished New Orleans in August 2005.

In a 1995 telephone conversation, the authors of this book speculated about what would happen if a Category 4 hurricane hit New Orleans. We forecast an abject disaster because the city lies below sea level, and such a storm would likely overwhelm the pumping system that must run to keep it dry. One thing we were sure of: No matter what the facts were, such a hurricane would be blamed on global warming. Ten years later, it happened.

On August 21, 2005, satellite imagery showed a very diffuse, but very *large* cloud mass beginning to organize east of the Bahamas. By the wee hours of August 24, the clouds had coalesced into a tropical depression, and by the evening of the 25th, just a few hours away from Miami, tropical storm Katrina became Category 1 Hurricane Katrina.

From the start, Katrina was an unusual tropical cyclone. It formed as a result of an addition of two separate tropical systems, a decaying tropical depression (a tropical cyclone with winds of less than 39 [miles per hour] mph) and a mass of thunderstorms that migrated across the Atlantic from Africa. As a result, Katrina was born with an unusually large cloud mass of showers and thundershowers.

The energy for a hurricane is derived from the condensation of water in the form of clouds. When matter goes from a less ordered (gaseous) state to a more ordered (liquid) one, heat is released to the surrounding environment. In a tropical cyclone, the center of the system becomes warmer than the surrounding environment. The more condensation, the stronger the system will eventually become. Katrina was born huge, primed to explode.

Florida got in the way. Right before hitting land, Katrina developed an eye, and had it not been for the nearby land, most forecasters think Katrina was on the edge of a catastrophic intensification, much like occurred in 1992, when Category 5 Hurricane Andrew had an extra 200 miles to run before hitting south Miami, compared to Katrina.

Katrina whacked a nation already hypersensitized to hurricanes. The previous year, 2004, was also a banner one, with 17 tropical cyclones. Four major storms affected Florida.

Hurricanes: From Really Bad to Impossibly Worse?

In chapter 7, we will discuss at length two questions that often arise in discussions about global warming among those who may not think it spells the end of the world. First, why is so little publicity given to scientific results consistent with that point of view, and, second, why does almost every finding seem to indicate that warming or its effects will likely be "even worse than we thought"?

One reason is that bad news sells. Even prestigious science periodicals such as *Nature* aren't beneath a bit of global warming embellishment.

Consider the photoshopped cover (see http://www.nature.com/nature/journal/v434/n7036/index.html) of its April 21, 2005, issue, which simultaneously displayed 2004 hurricanes Charley, Frances, Ivan, and Jeanne threatening Florida—a scientific impossibility. The actual dates of the storms were August 13, and September 3, 16, and 26.

On *Nature*'s cover, the hurricanes are in a physically unrealistic proximity. When hurricanes get too close together, one or both

Hurricane Intensity

The Saffir-Simpson scale of hurricane intensity classifies storms from Categories 1 (minimal) through 5 (extreme). Generally speaking, Category 1 and 2 storms are not particularly destructive, although there have been some notable exceptions, due to the potential for any tropical cyclone (including weak tropical storms that don't even make it to hurricane strength) to produce major flooding. In fact, many of the northeastern U.S. flood records are from Category 1 Hurricane Agnes in 1972.

Table 3.1
SAFFIR-SIMPSON SCALE OF HURRICANE INTENSITY

Category	Maximum One-Minute Average Wind
1	74–95 mph
2	96–110 mph
3	111–130 mph
4	131–155 mph
5	>155 mph

Category 3 or higher hurricanes are considered "major." Their frequency has changed over time, with low numbers in the 1930s and 1970s and high numbers in the 1950s and 1960s, and again since 1995. Roughly 40 percent of all hurricanes reach Category 3 at some point. Since 1900, 33 hurricanes have hit Category 5. Eight made landfall at that intensity somewhere in North or Central America, and three hit the United States. They were in 1935 (Florida Keys), 1969 (Camille in Mississippi), and 1992 (Andrew in South Florida).

storms will fall apart. That's because a healthy hurricane requires a large surrounding area aloft, called an outflow zone, in order to "vent" the rising air that forms the destructive vortex. When two storms are in relative proximity, the venting from one storm often destroys the venting from another. Four strong storms in such proximity has simply never happened because it can't happen.

The cover is explained on page ix of the issue:

> The 2004 hurricane season was one of the worst on record. Four hurricanes struck Florida in August and September. . . . On the cover (Courtesy University of Wisconsin–Madison, Space Science and Engineering Center) is a composite satellite image of hurricanes Charley, Francis, Ivan, and Jeanne "approaching" Florida in August and September 2004.

"Approaching"? I called University of Wisconsin in Madison to find out what was going on, and they replied that they had initially provided another image with the dates superimposed over each storm—a much less incendiary presentation, which *Nature* had declined as "too cluttered."

Ironically, immediately below the description of the cover is a reference to a "News Feature" article on page 952 titled, "Picture Imperfect":

> The magic of digital photography and Photoshop means that scientists can manipulate images so that key features are visible. But there is a grey area between image enhancement and misrepresentation. Helen Pearson reports ("News Feature," page 952).

Why bother separating the two paragraphs? They certainly would flow smoothly together. Or perhaps *Nature* should have said, "But there is a gray area between image enhancement and misrepresentation, *as shown on our cover.*"

So what *is* the science on global warming, hurricanes, and hurricane severity? Why are so many people convinced that they are increasing because of global warming?

Such one-sidedness is as near at hand as Al Gore's book and movie, *An Inconvenient Truth.* From the accompanying book, on pages 80–81:

> As the oceans get warmer, storms get stronger. In 2004, Florida was hit by four unusually powerful hurricanes. A growing number of new scientific studies are confirming that warmer water in the top layer of the ocean can drive more convection energy to fuel more powerful hurricanes. . . . But there is now a strong, new emerging consensus that global warming is indeed linked to a significant increase in both the duration and intensity of hurricanes.

In fact, buried within the scientific literature are a number of articles saying precisely the opposite. They might be fewer and further between than gloom-and-doom pieces, but they are there.

A good place to start this discussion is in the midst of the active 2004 season. On September 16 of that year, Thomas Knutson, of the National Oceanic and Atmospheric Administration, and Robert Tuleya published a paper in the *Journal of Climate* in which *computer-generated* hurricanes showed a slight increase in strength as carbon dioxide accumulated in the atmosphere.

New York Times science writer Andrew Revkin summarized their paper this way:

> Global warming is likely to produce a significant increase in the intensity and rainfall of hurricanes in coming decades, according to the most comprehensive computer analysis done so far.

That's not even close to what Knutson and Tuleya actually wrote! "CO_2-induced tropical cyclone intensity changes are unlikely to be detectable in historical observations," they concluded, "and will probably not be detectable for decades to come."

In the grand scheme of weather systems, hurricanes are actually pretty small and ephemeral—so small and short-lived that large-scale climate models that attempt to project global and regional temperature do not include them.

For that reason, Knutson and Tuleya began with model projections of future sea-surface temperatures (SSTs), vertical temperature profiles in the atmosphere, and vertical moisture profiles over regions where tropical cyclones form, using them to define a climate in which they used a finer-resolution hurricane model to spin up tropical cyclones. They then compared the characteristics of the *computer-generated* storms in the *computer-generated* future climate with the *computer-generated* storms in the current observed climate. They found that in the future climate, model-derived hurricanes had a 14 percent decrease in their lowest barometric pressure (a measure of intensity), a 6 percent increase in the maximum surface wind, and an 18 percent increase in the average rate of precipitation with 60 miles of the storm center over the model-derived hurricanes in the current climate. All these changes were indications that the model-derived hurricanes of the model-derived future would be more intense than the model-derived hurricanes of today.

Let's examine the modeled world that Knutson and Tuleya created and compare it with its real-world counterpart.

Carbon dioxide levels in the modeled atmosphere were increased at a rate of 1 percent per year. That rate leads to atmospheric carbon dioxide concentrations 80 years from now (the year that Knutson and Tuleya compared with current conditions) that are more than double the levels of today. It was at the end of this 80-year period when they calculated their changes in the storms.

Eighty years is a long time from now, but the actual time that these forecast changes would appear could be even further away.

That's because, in the real world, the concentration of atmospheric carbon dioxide has been growing at slightly less than half the rate used by Knutson and Tuleya. In the decade ending in 2004, the average increase was 0.49 percent per year (despite the rapid industrialization of China and India, the increase in the most recent year, 2006–07, remained at 0.49 percent), the decade before that 0.42 percent, and the one before that (1974–84), 0.48 percent. Obviously their model is based upon an overestimation of near-term carbon dioxide growth.

That has important implications. There is a lag time of several decades between changes in carbon dioxide and its final reflection in oceanic temperature. The process is somewhat analogous to the time it takes a pan of water to reach a constant temperature once a burner is turned on underneath it, although the physical processes for transferring heat are quite different between a shallow pan and a deep ocean.

Because the increase in atmospheric carbon dioxide in the real world isn't likely to reach 1 percent per year in the near future, that means that whatever changes in hurricanes are projected by Knutson and Tuleya for the next several decades have to be overestimations.

So let's be charitable and say that the 80-year changes they projected are reached in 100 years. By that time, it is highly likely that the energy structure of the world will be significantly different than it is today, possibly with fossil fuels being a curiosity of the past. Those model-presumed concentrations might never be reached.

The modeled hurricanes grow in a climate that is ideal. Specifically, there is virtually no change in wind speed or direction with height. That is called "wind shear," which basically blows the tops of the storms, preventing them from becoming well organized. One

El Niño = Los Weeños Hurricanes

El Niño ("the Child") is the oxymoronic name given to the biggest climate phenomenon on earth. It is a periodic disturbance that is so strong that it can reverse the largest wind system on the planet (the trade winds), and turn deserts into blooming gardens. We provided a brief description of its effects in chapter 1.

El Niño has a peculiar effect on hurricanes. The disturbance of the Eastern Pacific winds carries over into the Atlantic, where the midatmospheric winds acquire a much stronger westerly ("from the west") component than they would normally have. This creates a condition called "wind shear," in which the wind's velocity changes considerably with height.

Hurricanes can't stand wind shear. Hurricanes are a massive heat engine, with more and more air converged toward a central warm core, spun skyward, and whirled away in the outflow zone. If there's wind shear, the central circulation becomes distorted and—often—is completely blown apart, with the top of a former hurricane over a hundred miles away from the broken vortex at the surface, which remains visible until its clouds dissipate. So, El Niño turns hurricanes into los weeños.

When El Niño goes away, winds in the midatmosphere in the tropical Atlantic are more uniform, and there's usually a pretty good hurricane season.

phenomenon that is responsible for increasing the vertical wind shear in the tropical Atlantic is El Niño.

A number of studies have demonstrated that hurricane activity in the Atlantic Ocean decreases in years with El Niños, as well as the chance that those that do develop will hit the United States. Some climate models suggest increased El Niño–like conditions in the future; others don't. Knutson and Tuleya assumed that not only would there be no wind shear changes in the future, but that *there would be virtually no wind shear at all in any of their models.* This is the ideal climate for developing strong hurricanes—with the strength of the storms largely governed by the temperature of the underlying ocean surface.

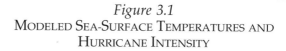

Figure 3.1
MODELED SEA-SURFACE TEMPERATURES AND HURRICANE INTENSITY

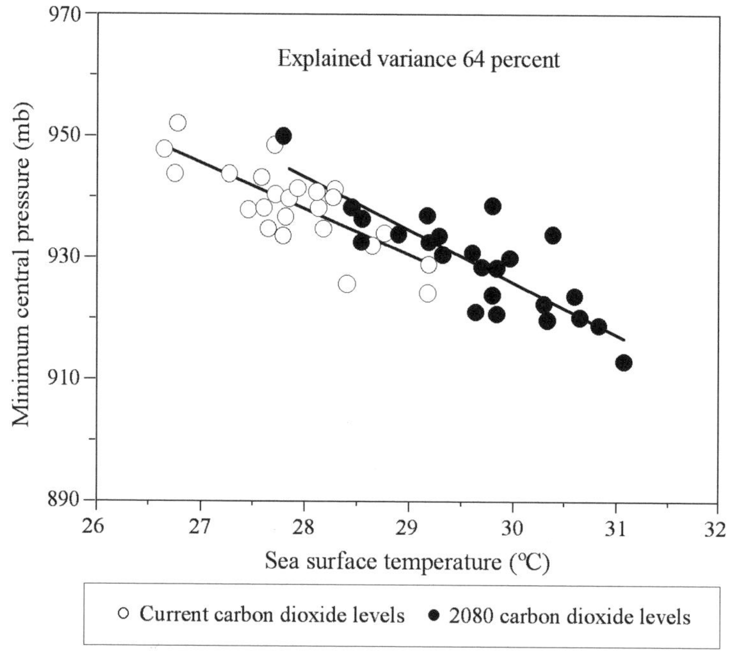

SOURCE: Knutson and Tuleya 2004.

NOTE: Storms with lower central pressures are generally stronger.

The authors, in fact, note a strong correlation between sea-surface temperatures and hurricane intensity—the warmer the sea surface, the stronger the storm. Figure 3.1 shows an example of the relationship between SSTs and hurricane intensity used by Knutson and Tuleya. In all their models, sea-surface temperatures alone explain an average of 55 percent of the changes in simulated hurricane intensity. The strength is measured by minimum central pressure (the lower the pressure, the stronger the storm). We show only one model run, with 64 percent explained variance.

Given that all the global climate models warm up the oceans when carbon dioxide levels are enhanced (even more so when the rate of carbon dioxide increase is larger than it actually is!), higher carbon

Explained Variance

What does it mean to say that sea-surface temperatures explained "an average of 55 percent of the changes in simulated hurricane activity?" That's the concept of "explained variance," or EV.

EV is a mathematical measure of the correspondence between two variables. If the EV between, say, sea-surface temperature and hurricane intensity were 100 percent, a plot of one vs. the other would correspond to a straight line or some easily simulated curve. As the explained variance falls, more and more points fall further and further off the line. When there's no explained variance, there's no line or uncomplicated curve that the points even appear to line up along.

In our hurricane example, the explained variance between computer-generated hurricanes and computer-generated sea-surface temperature was 64 percent, which is a pretty high number considering how many different factors can influence hurricane strength. In our study of real-world temperatures and SST, we found that EV was only 11 percent, or almost six times less. In other words, in reality, almost 90 percent of the behavior of hurricanes is determined by factors *other* than sea-surface temperature.

dioxide levels leads to higher SSTs, which lead to stronger tropical cyclones.

But the real world is not so kind to fledgling hurricanes. Though certainly the temperature of the underlying ocean surface is critical (the SST must be at least 26.7°C (80°F) for storms to even develop at all), other factors, such as wind shear, are as important.

Getting Real. . .

Maybe it would be a good idea to look at the relationship between sea-surface temperature and hurricane strength in the real world.

One of us (Michaels) looked at the number of major hurricanes (Category 3 or higher) vs. the departure from normal in seasonal SST back to 1950, and also the average peak wind speed in the five strongest storms in each year. Results are shown in Figure 3.2. These are two pretty reasonable measures of variation in hurricane strength from year to year.

The explained variance in this real-world analysis was far below what Knutson and Tuleya found in the virtual world: 11 percent (Michaels et al.) vs. 55 percent (the average of all the Knutson and Tuleya experiments). In other words, when all the factors that influence hurricanes are allowed to act, as must be the case when looking at real storms, the influence of SST drops by a factor of five.

More Powerful Storms?

Nine months later, two papers appeared in *Nature* and *Science* within a month of each other, both arguing that hurricanes are increasing in intensity.

The first, by MIT's Kerry Emanuel, reported a doubling of the "power" of hurricanes since the mid-1970s. Emanuel's mathematical index was based upon the third power (cube) of the hurricane maximum winds, as well as the frequency and lifetime of storms.

Emanuel reported that there was a significant correlation between the total power of storms in a year and the sea-surface temperature in the tropics, as well as the pattern of temperature departures from average in the North Atlantic and Pacific oceans.

Note the increase since in the number of major (Category 3 or higher) hurricanes since 1975, as shown in Figure 3.3. Now consult Figure 3.5 which is the Northern Hemisphere average surface temperature history from the United Nations. It, too, reaches a low point in 1975. So the earth warms, and the power of hurricanes increases.

Cause and effect or mere correlation? Anything—from hurricanes to the Consumer Price Index—that has increased since 1975 will obviously be correlated with global warming. But causation is much more elusive.

Emanuel's "power index" is largely determined by the cube of the total maximum winds in a given year. So if there is a linear rise (as has been observed) in the number of strong storms in recent decades, raising that to the third power gives a spectacular increase in his index.

Emanuel correlated his power index with three factors. He wrote: "I find that the record of net hurricane power dissipation is highly

Figure 3.2

OBSERVED RELATIONSHIP BETWEEN SEA-SURFACE TEMPERATURES AND NUMBER OF MAJOR HURRICANES (TOP) AND AVERAGE PEAK WIND SPEED IN THE FIVE STRONGEST YEARLY STORMS (BOTTOM)

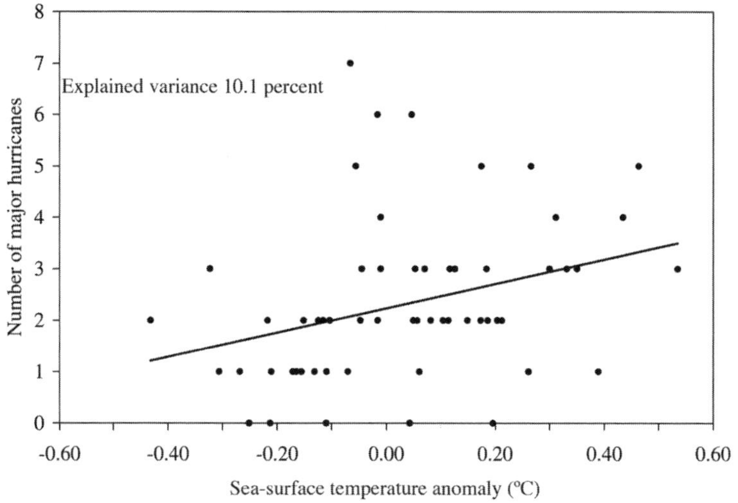

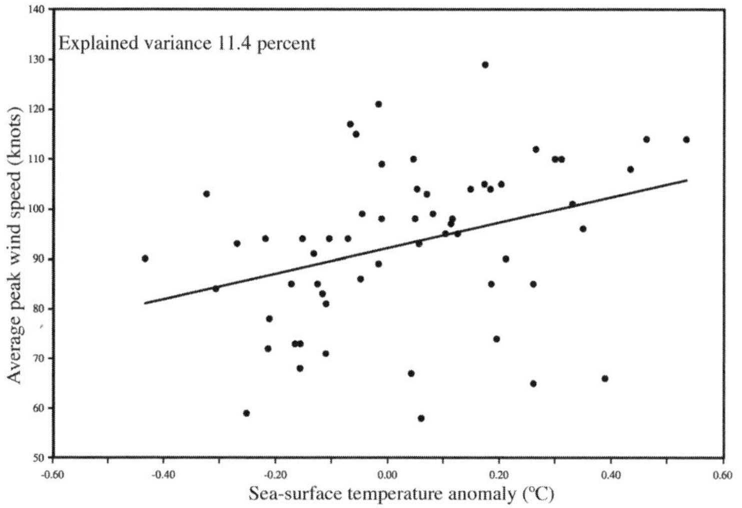

SOURCE: Michaels, Knappenberger, and Landsea 2005.

2005: The Biggest Year Ever?

The massive hurricane season of 2005—with 31 identified tropical storms and hurricanes—certainly got the public's attention. Not only were there a lot of storms, but there were also a lot of powerful ones, and some in pretty prominent places. Katrina, at the end of August, was a massively large cyclone. Though it made landfall in southeast Louisana as a "mere" Category 3 storm (down from a Category 5 with 170-mph sustained winds), its huge circulation piled a tremendous amount of water against the Mississippi and Alabama Gulf Coast, creating a storm surge that greatly exceeded the previous record-holder, 1969 Category 5 Hurricane Camille, which also hit the Mississippi coast.

Three weeks later, Hurricane Rita bombed out to Category 5, took a brief aim at Houston (where many of Katrina's evacuees were bivouacked), before also weakening to Category 3 and hitting the Texas–Louisana border. Then, in mid-October Hurricane Wilma literally exploded in the Western Caribbean, smashing all the previous records for speed of intensification and for lowest barometric pressure ever recorded in the Western Hemisphere, at 26.05 inches of mercury. Physically, that means that Wilma's intense cyclone blew away a full 12 percent of the atmosphere near its center. Wilma's winds also peaked at 170 mph.

There were also some pretty odd ones. In the far eastern Atlantic, north of the Canary Islands, minimal Hurricane Vince sprung up and made it to Spain as a tropical storm. Occasionally, weak tropical cyclones have gotten caught in strong westerly winds and hit Europe—the last time was in October 1992, when former hurricane Frances hit Spain. In the blogosphere, Vince's European landfall was touted as even more evidence for global warming, but in reality, it formed over unusually cold water and ameliorated a nasty drought in Portugal.

It's doubtless there have been numerous strikes like Vince, but before the days of satellites and hurricane-hunter aircraft, who would know some blustery rainstorm in France was in fact a former hurricane?

(continued on next page)

(continued)

Is 2005 truly the year with the largest number of tropical cyclones on record? Hard to say.

There were 21 identified storms in 1933, but no satellites or airplane sorties. Christopher Landsea, of the National Hurricane Center, has noted that many of the 2005 storms would have gone undetected back in the 1930s. In fact, examination of the actual tracks in 2005 indicates that as many as 12 would not have been reported in 1933, resulting in total of 33 storms in 1933 with today's detection technology (Figure 3.4; see insert). It is plausible that the 1933 total was in fact similar to the record number recorded in 2005.

Figure 3.3

NUMBER OF TROPICAL STORMS AND HURRICANES (LIGHT GRAY) AND MAJOR HURRICANES (DARK GRAY) (CATEGORIES 3, 4, AND 5), 1930–2007

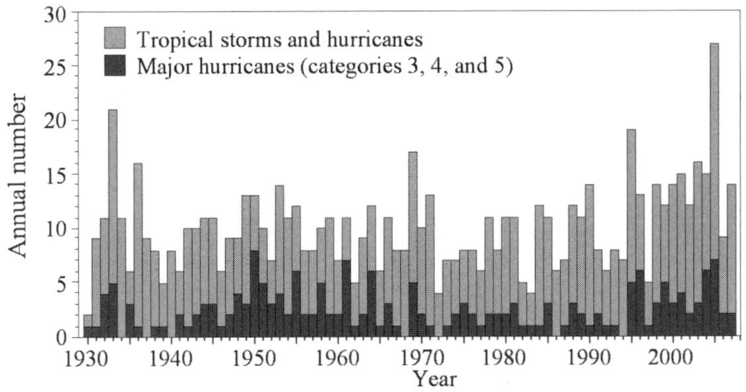

SOURCE: Unisys Weather 2008: http://weather.unisys.com/hurricane/atlantic/index.html.

correlated with tropical sea-surface temperature, reflecting well-documented climate signals, including multidecadal oscillations in the North Atlantic and North Pacific, and global warming."

Figure 3.5
NORTHERN HEMISPHERE AVERAGE SURFACE TEMPERATURES, 1930–2007

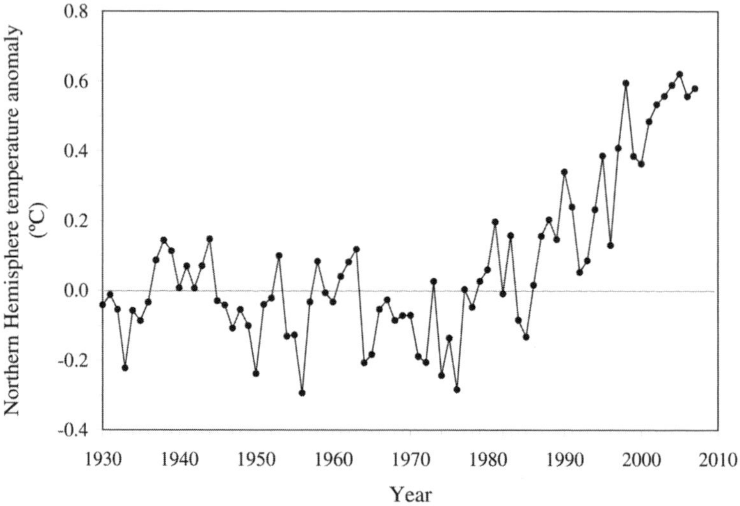

SOURCE: IPCC 2007 and updates.

Figure 3.6
THE ATLANTIC MULTIDECADAL OSCILLATION, 1870–1999

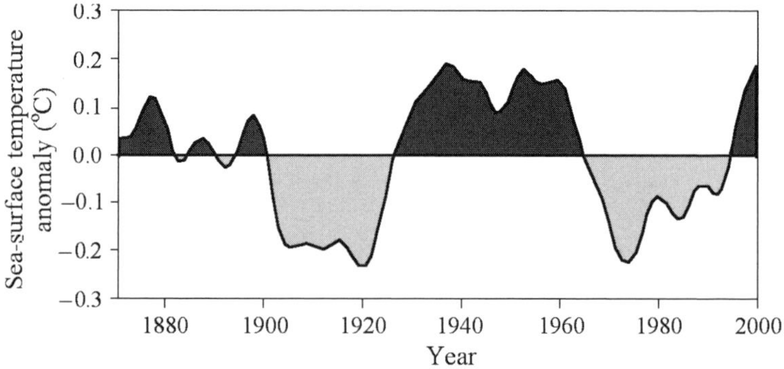

SOURCE: Knight et al. 2005.

The "oscillation" he is talking about in the Atlantic is known as the Atlantic Multidecadal Oscillation (AMO) and it has a history that predates global warming (Figure 3.6). The AMO is an index that reflects the temperature of the sea surface between Greenland and the Equator. Our figure shows it is hardly constant.

The AMO has long been associated with hurricanes. From the mid-1920s through the late 1960s, the AMO was in a warm state, and hurricane activity tended to be high. Then it went negative, and hurricanes decreased. Suddenly, in 1995, the AMO switched from a cold to a warm phase, and hurricane activity immediately increased, which is why some hurricane researchers, such as Christopher Landsea of the National Hurricane Center, believe the AMO's warm phase is what is primarily responsible for the recent uptick in hurricane activity. In 1995—as soon as he saw the AMO switch—Landsea actually predicted that hurricane activity was going to pick up, and that this was likely to continue for decades. His warning—which was not based on global warming—has proven correct.

What's the relationship between the AMO and global warming? Here's what the U.S. National Oceanic and Atmospheric Administration has to say:

> Instruments have observed AMO cycles only for the last 150 years, not long enough to conclusively answer this question. However, studies of paleoclimate proxies, such as tree rings and ice cores, have shown that oscillations similar to those observed instrumentally have been occurring for at least the last millennium. This is clearly longer than modern man has been affecting climate, so the AMO is probably a natural climate oscillation. In the 20th century, the climate swings of the AMO have alternately camouflaged and exaggerated the effects of global warming, and made attribution of global warming more difficult to ascertain. (http://www.aoml.noaa.gov/phod/amo_faq.php#faq_10)

If Emanuel is correct, damages should be mounting rapidly, given that a doubling of the strength of hurricanes would be exceedingly costly. The insured value of property from Brownsville, Texas, to Eastport, Maine—our hurricane-prone Atlantic Coast—is greater than a year of our Gross Domestic Product. If hurricanes had actually doubled in power, the financial effect would be catastrophic.

Roger Pielke Jr., from the University of Colorado in Boulder, has studied this subject, and his work is well known. Hurricanes are indeed causing greater dollar damages—because more and more people are building increasingly expensive beachfront monstrosities that have financially appreciated during the recent real-estate bubble. Account for those, and there is no significant change in hurricane expenses along our coast. Pielke told us that "analysis of hurricane damage over the past century shows no trend in hurricane destructiveness, once the data are adjusted to account for the dramatic growth along the nation's coasts."

Just a few weeks later, Peter Webster and colleagues, from Georgia Institute of Technology, published a paper in *Science* showing that, globally, since 1970, the number of tropical cyclones hasn't changed, but that the intensity has increased. They did find an increase in the number of storms in the Atlantic, but no change globally. That means that the frequency has to be in decline elsewhere. Hurricane numbers are going down in the North Pacific and Southern Hemisphere oceans. Unlike Emanuel, Webster specifically ruled out the warming of sea-surface temperatures as a cause:

> Only one region, the North Atlantic, shows a statistically significant increase, which commenced in 1995. However, a simple attribution of the increase in numbers of storms to a warming [sea-surface temperature] environment is not supported, because of the lack of a comparable correlation in other ocean basins where SST is also increasing.

That statement somehow didn't make the news reports. But what did get ink was the increase in intensity. Webster and his colleagues reported that the number of weak (Category 1) hurricanes had declined since 1970, that Categories 2 and 3 had shown no net change, and that the number of severe (Category 4 and 5) hurricanes had increased.

In the early 1970s, approximately 45 percent of all storms globally were Category 1. Category 2 and 3 storms contributed another 40 percent, and the severe Category 4 and 5 storms made up the remaining 15 percent. In the early 2000s, however, the annual contributions from those three groups was approximately equal. *That* made the news. Webster et al. had also reported that their result was "not inconsistent" with "recent climate model simulations."

Other models say otherwise. Masato Sugi, who heads up hurricane research at Japan's Meteorological Research Institute (a government entity), has run a global climate model to simulate hurricane behavior in a warming world, and found that tropical cyclone frequency *decreases* globally while there is no average change in intensity. The sum-total of that would be a *decrease* in destructive potential.

Lennart Bengtsson, of Germany's Max Planck Institut, has published multiple papers using computer models that project decreases in hurricane intensity or numbers. The first of those appeared in 1996.

More recently (and after Webster's publication), Akira Hasegawa of the Japan Agency for Marine-Earth Science and Technology simulated a decrease in both the intensity and the frequency of tropical cyclones in a warming world.

All in all, given that there are such conflicting studies on climate change and hurricanes, it would be fairer to say that Webster's finding is either consistent or inconsistent with recent climate model simulations.

Thanks to all the publicity, the UN's World Meteorological Organization issued a "Statement on Tropical Cyclones and Climate Change" in November 2006:

> During 2005 two highly publicized scientific papers appeared documenting evidence from the observational record for an increase in tropical cyclone activity. [The report then describes the Emanuel and Webster papers]. . . . Currently published theory and numerical modeling results suggest [a relatively small increase in tropical cyclone intensities several decades in the future], which is inconsistent with the observational studies of Emanuel (2005) and Webster et al. (2005) by a factor of 5 to 8 (for the Emanuel study) . . . this is still hotly debated area [sic] for which we can provide no definitive conclusion.

The problem with Webster et al. is the start date. Webster and his colleagues started in 1970 because that's the first year of satellite coverage. That's also very close to the start time for the warming that has been observed since 1975. That makes for a lot of correlation, but not a lot of causation.

Pielke and four very prominent coauthors[1] published a much different study in the *Bulletin of the American Meteorological Society* in 2005. Here's their major conclusion:

[1] Pielke's coauthors were Christopher Landsea, a leading researcher on hurricanes and climate; Max Mayfield, former Director of the National Hurricane Center; James Laver, head of the federal Climate Prediction Center; and Richard Pasch, hurricane specialist at the National Hurricane Center.

> To summarize, claims of linkages between global warming and hurricane impacts are premature for three reasons. First, no connection has been established between greenhouse gas emissions and the observed behavior of hurricanes (Houghton et al. 2001; Walsh 2004). Emanuel (2005) is suggestive of such a connection, but is by no means definitive. In the future, such a connection may be established [e.g., in the case of the observations of Emanuel (2005) or the projections of Knutson and Tuleya (2004)] or made in the context of other metrics of tropical cyclone intensity and duration that remain to be closely examined. Second, the peer-reviewed literature reflects that a scientific consensus exists that any future changes in hurricane intensities will likely be small in the context of observed variability (Knutson and Tuleya 2004; Henderson-Sellers et al. 1998), while the scientific problem of tropical cyclogenesis [formation] is so far from being solved that little can be said about possible changes in frequency. And third, under the assumptions of the IPCC, expected future damages to society of its projected changes in the behavior of hurricanes are dwarfed by the influence of its own projections of growing wealth and population (Pielke et al. 2000). While future research or experience may yet overturn these conclusions, the state of the peer-reviewed knowledge today is such that there are good reasons to expect that any conclusive connection between global warming and hurricanes or their impacts will not be made in the near term.

Pielke, a professor of environmental studies at University of Colorado, is no political neophyte. He worked for the late Congressman George Brown (D-CA), the powerful chair of the House Science Committee. He's a self-described "Blue-Dog Democrat," and he has written extensively on the interactions among science, policy, scientists, and society. His website "Prometheus: the Science Policy Weblog" (http://sciencpolicy.colorado.edu/prometheus) probably has the best discussions in cyberspace on this nexus.

At any rate, Pielke et al. were not reluctant to share their feelings about the political misuse of hurricanes and global warming. Here's an excerpt from the next paragraph in their paper:

> Yet claims of such connections persist (cf. Epstein and McCarthy 2004; Eilperin 2005), particularly in support of a political agenda focused on greenhouse gas emissions reduction (e.g., Harvard Medical School 2004). But a great irony here is that invoking the modulation of future hurricanes to justify

> energy policies to mitigate climate change may prove counterproductive. Not only does this provide a great opening for criticism of the underlying scientific reasoning, it leads to advocacy of policies that simply will not be effective with respect to addressing future hurricane impacts. There are much, much better ways to deal with the threat of hurricanes than with energy policies (e.g., Pielke and Pielke 1997). There are also much, much better ways to justify climate mitigation policies than with hurricanes (e.g., Rayner 2004).

Pielke's paper obviously ruffled some academic egos. Even before it was published, Kevin Trenberth from the National Center for Atmospheric Research (also in Boulder) told the local newspaper, "I think [Pielke] should withdraw his article. This is a shameful article."

"Shameful"? In fact, Pielke's logic is quite sound. If there were a strong link between global warming and hurricanes, which costs more? Adaptation to them, or a futile attempt to stop warming? The latter takes away resources from the former, while accomplishing nothing.

Dozens of news stories about the hyperactive 2005 hurricane season mentioned that ocean temperatures in the Atlantic Basin were very warm that year, and, by implication, global warming juiced up the monster storms. Is that really the case?

In spring of 2006, one of us (Michaels) published a paper in *Geophysical Research Letters* showing no relationship between the maximum sea-surface temperature over which a hurricane passes and that maximum winds observed in the life of strong hurricanes. All storms were studied that experienced waters of 28.25°C (82.9°F), the threshold that is required for Category 3 ("major") hurricanes. Temperatures in the Gulf of Mexico, and much of the southwestern Atlantic Ocean, exceed this value for many months every year.

Kerry Emanuel responded that we hadn't looked at enough hurricanes. In fact, we had looked at all storms that experienced this warm water since 1982, the year that an appropriate record of ocean temperatures begins. There were 195, an average of about 9 a year. Emanuel was able to generate a significant relationship by manufacturing 3,000 computer-generated hurricanes.

Philip Klotzbach of Colorado State University then published a paper in *Geophysical Research Letters* that examined "worldwide tropical cyclone frequency and intensity to determine trends in activity over the past 20 years during which there has been an approximate

Living with Hurricanes

On September 28, 1955, a Category 5 hurricane named Janet slammed into Chetumal, on Mexico's Yucatan Peninsula, killing more than 600 people.

On August 21, 2007, Hurricane Dean, another Category 5 and the third-strongest storm ever measured at landfall, hit within a few miles of where Janet struck, and killed no one. (Mountain flooding from the storm's remnants resulted in eight fatalities well inland). Maximum winds in Janet and Dean were the same. Dean most likely marked the first instance in human history in which a Category 5 hurricane hit a populated coast and everyone lived.

In 2005, Hurricane Wilma, a Category 4 storm (the same intensity of cyclone that killed 7,000 people in Galveston, Texas, in 1900), hit the tourist-heavy northeast corner of the Yucatan and killed only four people.

Because of its peculiar location, the Yucatan takes more big hurricane hits than just about anywhere else in the Western Hemisphere. When Mexico was dirt-poor, as it was in 1955, hurricanes could kill hundreds. They were warned then, too. Hurricane-hunter planes also monitored Janet. Only one of those has ever been lost, and it was as Janet was making landfall.

Similar storms. Huge storms. Very different results. What changed?

Prior to the development of tropical meteorology in the mid-20th century, storms such as these used to kill hundreds, even thousands, as they zeroed in on unsuspecting populations. But we now have the technology to forecast their tracks, at least for the critical last 48 hours, with reasonable confidence. That gives people time to evacuate. Economic development gives people the infrastructure necessary to accommodate evacuation. When Janet killed hundreds, per capita income in Mexico was less than a tenth of what it is now.

Will global warming change this? Note that Knutson and Tuleya calculated that maximum winds should increase by about 6 percent over the next 75 years. Even that may be

(continued on next page)

(continued)

an overestimate because they assume that carbon dioxide is increasing in the atmosphere about twice as fast as it actually is.

Clearly, this small increase in hurricane strength is going to be dramatically overshadowed by adaptation as the developing world continues to develop. Mexico is a case in point.

Anyone concerned about climate change should take a lesson from Hurricane Dean. Even if storms like that one become more frequent in the future, people will adapt and survive—provided they have sufficient financial resources. How silly it seems to take resources away in futile attempts to "stop global warming" when those same resources can be directed toward adaptation, including infrastructure and hurricane-proof housing.

The truth is that money in the hand is a lot more useful than treaties on paper when it comes to adapting to severe weather. So people truly worried about climate change should be cheerleading for economic development, which provides the resources necessary to accommodate even the strongest hurricanes.

0.2°C to 0.4°C [0.4°F to 0.7°F] warming of SSTs." Klotzbach found "a large increasing trend in tropical cyclone intensity and longevity for the North Atlantic basin and a considerable decreasing trend for the North Pacific." The increase in the Atlantic was exactly the same as the decrease observed in the Northeast Pacific ocean (Figure 3.7). Other tropical cyclone-producing ocean basins showed only small variations.

Overall, Klotzbach noted "no significant change in global net tropical cyclone activity" but a "small increase in global Category 4–5 hurricanes from the period 1986–95 to the period 1996–2005." His metric was "Accumulated Cyclone Energy" (ACE), an integrated measure of total storm strength in a year. From this analysis, he concluded that factors other than sea-surface temperatures are important in governing tropical cyclone frequency and intensity and noted the likelihood that "improved observational technology" has also had an influence on the small increases that he did observe.

Figure 3.7
ACCUMULATED CYCLONE ENERGY (ACE) INDEX FOR THE WORLD'S HURRICANE BASINS, 1986–2005

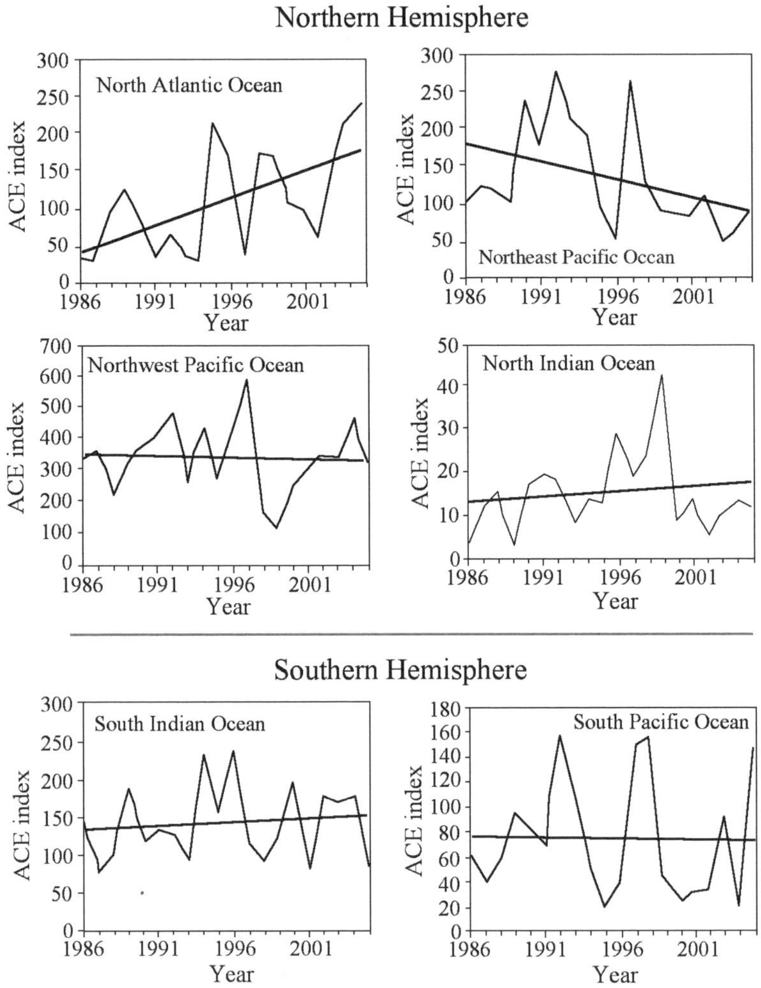

SOURCE: Klotzbach and Gray 2006.

Klotzbach ultimately summed up his findings as

> . . . contradictory to the conclusions drawn by Emanuel (2005) and Webster et al. (2005). They do not support the argument

> that global [tropical cyclone] frequency, intensity, and longevity have undergone increases in recent years. Utilizing global "best track" data, there has been no significant increasing trend in ACE and only a small increase (~10 percent) in Category 4–5 hurricanes over the past 20 years, despite an increase in the trend of warming sea-surface temperatures during this time period.

Two more major hurricane papers followed in fall 2006. First, Klotzbach (along with Colorado State University's William Gray) examined the very destructive 2004 season.

They noted, as Klotzbach had in his earlier paper, that there was an increase in activity in the Atlantic beginning in 1995, but that there were equivalent or greater declines in the rest of the world. They attributed those findings to changes in the *distribution* of temperature in the Atlantic, rather than to global warming. They also echoed the conclusions of Pielke et al.: "Due to increased coastal population and wealth, the U.S. coastline can expect hurricane-spawned damage and destruction in the coming few decades to be on a scale much greater than has occurred in the past."

Chris Landsea et al. then weighed in, in *Science*, asking whether the data for tropical cyclones are in fact reliable enough to be used to detect long-term trends.

Before landfall, hurricane intensities are either measured by hurricane-hunter aircraft or by satellite. Only two regions, the western Atlantic and western Pacific, have had regular aircraft reconnaissance, which provides a fairly homogenous set of data back to at least 1960. Satellite monitoring began in 1970, but the onboard instrumentation has improved over time, with more recent orbiters able to provide higher-resolution images and direct views of storms that allow for more accurate estimates of highest winds. So, any histories that are primarily satellite-based are likely to show an artificial upward trend in intensity. The aircraft-based histories show no significant trends at all. Landsea and his colleagues concluded "that extreme tropical cyclones and overall tropical cyclone activity have globally been flat from 1986 until 2005, despite a sea-surface temperature warming of 0.25°C [0.45°F]."

The more you look, the less obvious it becomes that anthropogenic global warming has significantly (i.e., measurably) contributed to the current increase in hurricane activity in the North Atlantic basin, or anywhere else in the world, for that matter.

270 Years of Hurricane History!

Sounds like the same-old same-old. An article appeared in a 2007 issue of *Nature* with the first sentence in the abstract stating, "Hurricane activity in the North Atlantic Ocean has increased significantly since 1995." One can only guess the dire global warming news to follow.

Johan Nyberg, from the Geological Survey of Sweden, and his coauthors began with this:

> The years from 1995 to 2005 experienced an average of 4.1 major Atlantic hurricanes (Category 3 to 5) per year, while the years 1971 to 1994 experienced an average of 1.5 major hurricanes per year. This increase in major hurricane frequency is thought to be caused by weaker vertical wind shear [the strength of winds with height] and warmer sea-surface temperatures (SSTs) in the tropical and subtropical Atlantic.

The title of the Nyberg et al. article, "Low Atlantic Hurricane Activity in the 1970s and 1980s Compared to the Past 270 Years," indicates that hurricane activity was low in the 1970s and 1980s compared with the past 270 years. Could it be that what we are seeing now is actually a return to more normal conditions?

How does one get a 270-year history of hurricanes? Corals growing in the Caribbean Sea preserve a year-to-year luminescence intensity (something like color differences), and as noted by Nyberg et al., "luminescence intensity in corals reflects the degree of terrestrial water runoff, a result of low precipitation, which is highly correlated with low hurricane activity." Nyberg and colleagues also examined plankton from a sediment core from the Caribbean. Certain plankton are associated with weaker hurricane regimes. The deeper the plankton are buried, the older they are. They presented convincing evidence that the corals and plankton accurately reflect hurricane activity during the period of reliable records (Figure 3.8). Their conclusion:

> The record indicates that the average frequency of major hurricanes decreased gradually from the 1760s until the early 1990s, reaching anomalously low values during the 1970s and 1980s. Furthermore, the phase of enhanced hurricane activity since 1995 is not unusual compared to other periods of high hurricane activity in the record and thus appears to

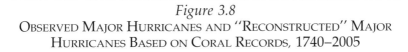

Figure 3.8
OBSERVED MAJOR HURRICANES AND "RECONSTRUCTED" MAJOR HURRICANES BASED ON CORAL RECORDS, 1740–2005

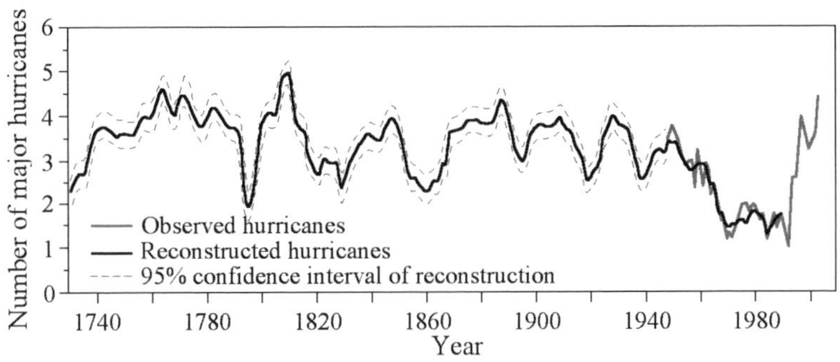

SOURCE: Adapted from Nyberg et al. 2007.

NOTE: The dashed lines are the 95 percent confidence intervals about each reconstructed value.

> represent a recovery to normal hurricane activity, rather than a direct response to increasing sea-surface temperature.

More specifically, they note:

> Only the periods ~1730–1736, 1793–1799, 1827–1830, 1852–1866 and 1915–1926 appear to have been marked by similarly low major hurricane activity. . . . Furthermore, the current active phase (1995–2005) is unexceptional compared to the other high-activity periods of ~1756–1774, 1780–1785, 1801–1812, 1840–1850, 1873–1890, and 1928–1933, and appears to represent a recovery to normal hurricane activity, despite the increase in SST.

Instead of being unusually active, it looks like the current hurricane regime is simply a return to more normal conditions, following an unusually tranquil couple of decades.

800 Years of Hurricane History!

Another long-term record of hurricanes comes from Australia (hurricanes are called "tropical cyclones" in Australia), where their passages are recorded in caves.

Stalagmites growing upward in caves can be "dated" with each year's rainy season. Jonathan Nott of Australia's James Cook University used them to create a record of tropical cyclone activity in northeastern Australia for the last 800 years. He published his record in a 2007 edition of *Earth and Planetary Science Letters.*

Let's pause for a science lesson: Water can contain two different isotopes of oxygen. Almost all the oxygen in water has a molecular weight of 16 (8 protons, 8 neutrons, remember?). But a very small fraction incorporates Oxygen-18, which contains two extra neutrons.

What does this have to do with hurricanes? Hurricanes are a major source of very dense, high cloudiness in that part of Australia, and rain that forms at high altitude contains very little Oxygen-18. Years in which there is very little of that isotope incorporated in a stalagmite will likely be years in which there was major hurricane activity.

A cave in nearby Chillagoe (see map, Figure 3.9) is full of upward-growing stalagmites, and the water from tropical cyclones contributes to the growth of each year's layer.

The Australian Meteorological Office has kept an excellent observational record of tropical cyclone activity in the region from 1907 to 2003. Consequently, they could compare yearly Oxygen-18 values in the stalagmites with the actual hurricane frequency. The researchers found that each peak in the depletion of this isotope "corresponds to the passage of a cyclone within 400 km (250 miles) of Chillagoe." There were 27 such storms in the record, and many passed much more closely. Those storms accounted for 63 percent of all the hurricanes that passed within 200 km (125 miles) of the cave.

Nott et al. noted:

> Despite the absence of many cyclones, it is important to note that every intense cyclone (i.e., AD 1911, 1918, 1925, 1934, 1986 as determined by barometer or damage to urban infrastructure and loss of life) to make landfall in the [400 km region] since AD 1907 is registered by a peak in the isotope depletion curve.

This is an amazing history. Figure 3.10 shows the tropical cyclone record from the Chillagoe cave. Note that the current era (1800 to present) is pretty wimpy in the broad historical sweep. The solid black line in Figure 3.10 is the threshold for an extreme storm, matching the great 1911 cyclone. Note that it is the only event of

Figure 3.9
NORTHEASTERN AUSTRALIA, THE LOCATION OF NOTT'S CAVE STUDY SITE

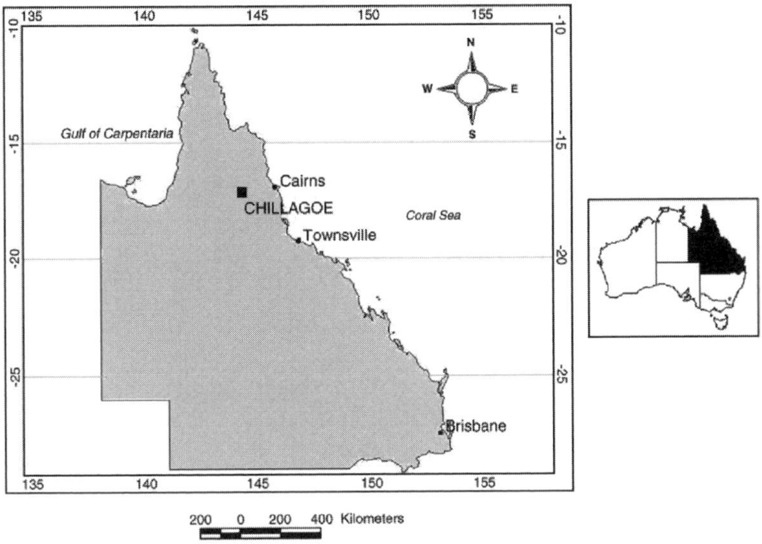

SOURCE: Nott et al. 2007.
NOTE: 1 kilometer = 0.62 miles.

such magnitude in the past 200 years, but that there were seven such storms in the previous two centuries.

5,000 Years of Hurricane History!

The marine forest at Vieques, a few miles east of Puerto Rico, has a nice tropical beach backed by a vegetated barrier ridge about 10 feet tall. Behind the ridge is a back-barrier lagoon that over time became a "playa," which is a flat-bottomed feature that is occasionally covered with water—such as when hurricanes flood it. During large hurricanes, which are fairly common in that area, the ridge is breached and a large amount of material is deposited on the playa. Brown University's Jeffrey Donnelly and Woods Hole Oceanographic Institution's J. D. Woodruff extracted cores from the playa, noting:

Figure 3.10
STRENGTH INDEX OF TROPICAL CYCLONE EVENTS, 1226–2003

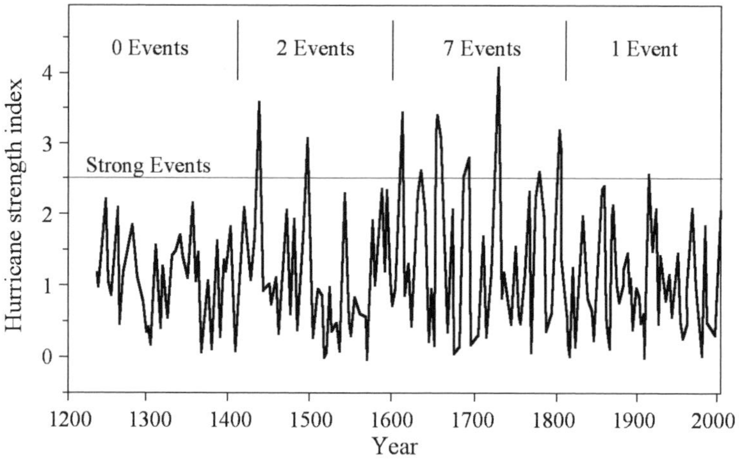

SOURCE: Adapted from Nott et al. 2007.

> Cores collected from the site contain several metres of organic-rich silt interbedded with coarse-grained event layers comprised of a mixture of siliciclastic sand and calcium carbonate shells and shell fragments. These layers are the result of marine flooding events overtopping or breaching the barrier and transporting these barrier and nearshore sediments into the lagoon.

Organic material can be dated with commonly used techniques (such as carbon dating, which has been employed for decades), and just like magic, a long-term record of intense hurricane activity is produced.

Donnelly and Woodruff reported:

> On the basis of our age model an interval of relatively frequent intense hurricane strikes at Vieques is evident between 5,400 and 3,600 calendar years before present ("yr BP," where present is defined as AD 1950 by convention), with the exception of a short-lived quiescent interval between approximately 4,900 and 5,050 yr BP. Following this relatively active period is an interval of relatively few extreme coastal flooding events persisting from 3,600 until roughly 2,500 yr BP.

> Evidence of another relatively active interval of intense hurricane strikes is evident between 2,500 and approximately 1,000 yr BP. The interval from 1,000 to 250 yr BP was relatively quiescent with evidence of only one prominent event occurring around 500 yr BP. A relatively active regime has resumed since about 250 yr BP (1700 AD).

With respect to the linkage between higher sea-surface temperatures and hurricane activity, the pair notes:

> Given the increase of intense hurricane landfalls during the later half of the Little Ice Age (around AD 1700), tropical SSTs as warm as at present are apparently not a requisite condition for increased intense hurricane activity. In addition, the Caribbean experienced a relatively active interval of intense hurricanes for more than a millennium when local SSTs were on average cooler than modern.

The authors obviously can't finger global warming as the cause, but instead cite variations in El Niño and African disturbances that for years have been associated with hurricane frequency.

Should another unusually intense hurricane season (such as 2005) spin up in the Atlantic, global warming advocates will be in front of every camera in sight claiming we are witnessing yet another manifestation of global warming. Why, then, was there more hurricane activity when the planet was cooler?

Endnote: Hurricanes in the Big Apple!

We started this chapter with Katrina in New Orleans. But if a large hurricane strikes New York City, the amount of devastation could be incredible, as well as the hue and cry blaming global warming. Perhaps this will be the $500 billion hurricane. (The costliest storm, assuming today's property values and population, was the 1926 hurricane that struck southeast Florida and Alabama, at $164 billion) But could a hurricane really devastate the Big Apple?

Indeed it could. The New York area has been struck many times in the past by tropical cyclones, so it's just a matter of time before another one passes directly over the city.

A recent article in *Geochemistry, Geophysics, Geosystems* by geological scientists at Brown University and Woods Hole Oceanographic

Figure 3.11
STORM SURGE HEIGHTS, 1788 TO PRESENT

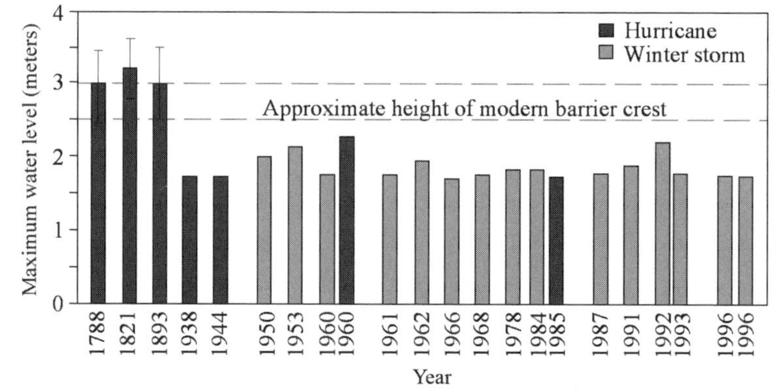

SOURCE: Adapted from Scileppi and Donnelly 2007.

NOTE: Storm surge heights relative to the modern mean sea level that accompanied the 1788, 1821, and 1893 hurricanes are inferred from historic archives and records of the most extreme flooding events of the 20th century recorded by the Battery Park, New York City, tide gauge from 1920 to present.

Institution focuses on hurricanes in the New York City area—specifically western Long Island. Elyse Scileppi and the aforementioned Jeffrey Donnelly begin their article by noting:

> Historical records show that New York City is at risk of being struck by a hurricane. Four documented strong hurricanes (Category 2 or higher on the Saffir-Simpson Scale) with high storm surges (~3 m) [10 feet] have made landfall in the New York City area since 1693 with the last occurring in 1893. Population growth during the 20th century has significantly increased the risk to lives and property should a strong hurricane recur today. The frequency of hurricane landfalls is difficult to estimate from the instrumental and documentary records due to the relative rarity of these events and the short historical observation period.

Hurricanes scour Long Island's beaches, and their storm surges then deposit sand inland, in muddy marshlands. The sand layers are pretty obvious, and they can be dated using a variety of methods.

The largest inundations were in 1788, 1821, and 1893 (Figure 3.11). The three large surges occurred during a time when the Northern Hemisphere was considerably cooler than it is now.

Figure 3.12
TRACKS OF FIVE HURRICANES AFFECTING THE WESTERN LONG ISLAND, NEW YORK, REGION, 1788–1985

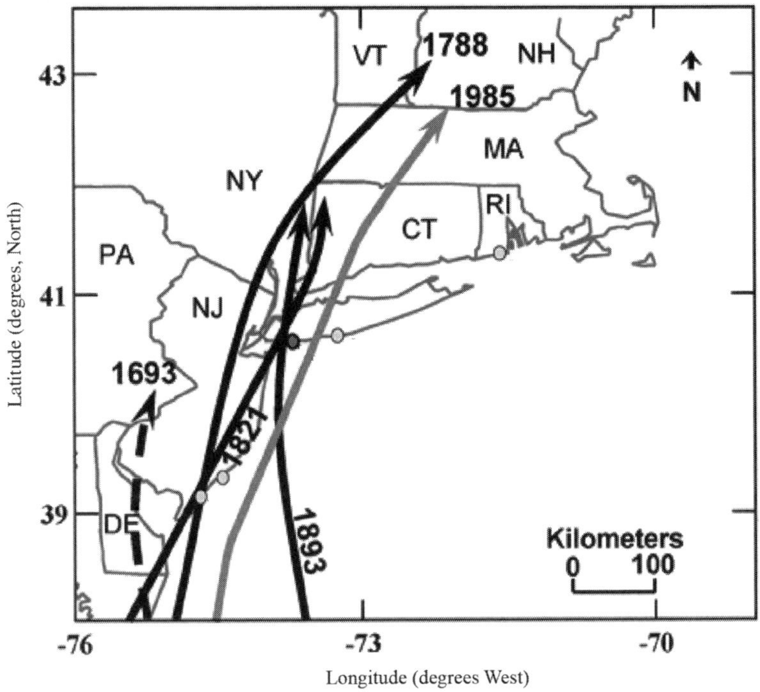

SOURCE: Adapted from Scileppi and Donnelly 2007.

NOTE: The dark tracks (1788, 1821, and 1893) indicate full-impact hurricanes. The gray track (1985) indicates a near-miss—in this case, Hurricane Gloria. The dashed track (1693) indicates uncertainty about the hurricane's path.

In terms of actual hurricane activity (Figure 3.12), they state: "Four historically documented hurricanes that caused approximately 3 m [10 feet] of storm surge made landfall in the New York City area in 1893, 1821, 1788, and likely 1693." They correctly note that things were quite a bit cooler than today:

> Interestingly, several major hurricanes occur in the western Long Island record during the latter part of the Little Ice Age (~1550–1850 A.D.) when SSTs were generally colder than present. According to paleoclimate estimates, SSTs were

> likely 2°C cooler than present in the Caribbean, 1°C cooler than present in the Florida Keys during the latter part of the Little Ice Age, and 1°C cooler than present during the 17th and 18th centuries at the Bermuda Rise.

The authors note, "Despite significantly cooler than modern [sea-surface temperatures] in the Atlantic during the latter half of the Little Ice Age, the frequency of intense hurricane landfalls increased during this time."

If they had found increasing New York hurricanes, they would have been paraded right down Broadway. But finding increased activity in colder periods certainly rained on *that* parade!

Summing Up Hurricanes

The hurricane–global warming link is clearly much more complicated than the simple "warming in—bigger hurricanes out" stories that abound today. It's not at all clear that a warmer world will have more storms, and many studies indicate that increases in hurricane strength will be hard to detect because of the tremendous year-to-year natural variability. We don't even have decent hurricane histories, because satellite coverage—the only way to truly measure global activity—has only been maintained for 35 years. Aircraft have investigated storms in the Atlantic and Western Pacific since World War II, but coverage was certainly not complete, and may not be reliable before 1960. Long-term geological records for individual sites can be used to find evidence for storms as far back as 5,000 years ago, and those histories indicate that there is nothing really unusual about the current hurricane regime.

That's all a far cry from the current noise about tropical cyclones and global warming. Although these scientific findings are there for all to see in the refereed literature, they certainly have received a lot less coverage then their more gloomy counterparts.

4. Sea-Level Rise and the Great Unfreezing World

Warning: You're going to read about data sets "guaranteed" to show large losses of Arctic ice, a newly discovered "island" uncovered by global warming (that was actually an island a mere half-century ago), and a scientific "urban legend" that almost all of Greenland's ice is going to crash into the sea, pronto.

Horrifying images of Greenland's ice sliding into the sea and raising the sea-level 10 or more feet by the year 2100 are common now. They owe their viability to one NASA scientist: James Hansen.

Hansen's Scenario

James E. Hansen, director of the NASA Goddard Institute for Space Studies, is the clear progenitor of the modern apocalyptic theory of climate change. His disaster hypothesis first appeared in 2004 in *Scientific American* (which does not vet its articles via the peer review process that academic journals use), and then he swore by it in two legal proceedings.

Hansen was involved in two important court cases in 2006–07. The first was in California, in which "Central Valley Chrysler-Jeep" sued the state of California, claiming that regulations for carbon dioxide emissions promulgated by the California Air Resources Board (at the direction of the California Legislature) would be impossible to meet and would result in their bankruptcy. The Vermont legislature then agreed to do everything that California would do, so there was another suit in Vermont. The California judge continued the case until *Massachusetts v. EPA* (see chapter 7) was settled by the U.S. Supreme Court; but the Vermont case went forward.

Hansen testified that if there were even less than 1°C [1.8°F] of post-2000 warming, then there was the "possibility of initiating ice sheet response that begins to run out of control" with ultimately "several" meters of sea-level rise. If warming proceeded according to the midrange estimate for carbon dioxide changes (Figure 1.5; see insert), Hansen stated that a sea-level rise of 6 meters (roughly 20 feet) by 2100 would be within the confidence limits of his estimate.

Note how far Hansen's predictions are from those of the Intergovernmental Panel on Climate Change, which says that the Greenland contribution to sea-level rise by 2100 is likely to be around two *inches*. (The IPCC puts in a small caveat that their estimate does not take into account changes in the ice that have not been modeled.)

Hansen even has an explanation for why climate scientists don't support his position. In a 2007 essay in *Environmental Research Letters*, Hansen claimed the reason for this was something he called "scientific reticence," or the desire of scientists to *not* publish or speak of bad news. As will be seen in chapter 7, quite the opposite is true.

At any rate, Hansen is by far the most quoted climate researcher in the world on Greenland (Google it, and you'll get about 80,000 hits for "James Hansen + Greenland"), despite his repeated protestations that he is being prevented from speaking out.

The Greenland myth became video in Al Gore's *An Inconvenient Truth*, in which he shows a montage of Florida being slowly submerged as Greenland loses its ice. Is that truth or fiction?

The IPCC's 2007 "Fourth Assessment Report" projects sea-level rise of between 8.5 and 18.5 inches for the 21st century for its "midrange" estimate of carbon dioxide and other greenhouse gas emissions.

At the top end, that represents a 32 percent reduction in estimated sea-level rise for the century (down from 27 inches) from its "Third Assessment Report," published in 2001. The mean, or central, value is 13.5 inches.

Of that amount, 66 percent of the rise, or 8.8 inches, is from expansion of warm water. That is directly proportional to the expected temperature rise in global temperature. The midrange emission scenario (Figure 1.5; see insert) results in an average modeled warming of approximately 4.9°F (2.7°C) between 2000 and 2100. If, as we argued in chapter 1, the warming is likely to be less, around 3.2°F (1.75°C), then the sea-level rise from thermal expansion will drop proportionally, to about 5.7 inches.

Another Perspective

In the "Policymaker's Summary" of its 2007 science compendium, the IPCC states:

> Global average sea-level rose at an average rate of 1.8 [1.3 to 2.3] mm per year over 1961 to 2003. The rate was faster

> over 1993 to 2003: about 3.1 [2.4 to 3.8] mm per year. Whether the faster rate for 1993 to 2003 reflects decadal variability or an increase in the longer-term trend is unclear.

One problem with science compendia such as the IPCC reports is that they must have discrete "cut-off" dates beyond which they include no published science. Otherwise, reports would be in a continuous state of revision.

It's too bad. A 2007 article by G. B. Wöppelmann and others published in *Global and Planetary Change* was beyond the cut-off, and could have changed the IPCC's speculation that the rate of sea-level rise may be increasing.

Measuring sea-level rise is far from simple. One main reason is the (geologically) recent ice age. The enormous ice sheets that covered much of our hemisphere pushed down on the crust, and the recovery process is slow. In fact, the crust is still rising. Scientists attempt to account for this effect using numerical "Glacial-Isostatic Adjustment" routines in their estimates of true sea-level rise. But movements of the continental plates, wind and ocean currents, and differing magnitudes of gravity also confound measurement of true sea level.

Wöppelmann et al. note that

> two important problems arise when using tide gauges to estimate the rate of global sea-level rise. The first is the fact that tide gauges measure sea level relative to a point attached to the land which can move vertically at rates comparable to the long-term sea-level signal. The second problem is the spatial distribution of the tide gauges, in particular those with long records, which are restricted to the coastlines.

Chances are that you now have a global positioning systems (GPS) unit in your car or boat, a GPS upgrade for your cell phone, or a hand-held GPS unit for hiking. GPS satellites are taking measurements of anything and everything, and data from advanced GPS networks now resolve questions about sea-level rise. Noting this new source of objective data, Wöppelmann et al. analyzed 224 GPS stations; 160 were located within 15 km [9.3 miles] of a tide gauge station (Figure 4.1). The data allowed them to very accurately measure the vertical motion of the crust from January 1999 to August 2005, and although the 7.7-year time span would seem rather short, they effectively argue that vertical motion of the crust is not like the weather—the vertical motion remains the same over long periods of time.

Figure 4.1
DISTRIBUTION OF 224 GPS STATIONS PROCESSED BY WÖPPELMANN ET AL., 2007

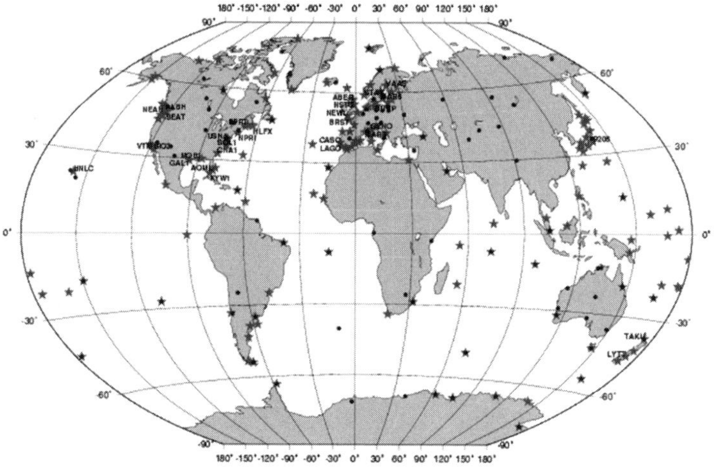

SOURCE: Wöppelmann et al. 2007.

NOTE: Stars are GPS stations less than 15 km (9.3 miles) from a tide gauge. Dots are continental stations.

When Wöppelmann et al. factored their measurements of land motion into the estimate of sea-level rise, they determined a global value of 1.31 ± 0.30 mm per year (0.05 ± 0.01 inches) compared with the 3.1 mm value given for recent years by the UN.

Where's the headline? "Objective Measurements Reduce Recent Sea-Level Rise by Nearly 70 Percent!" Of course, this book is about the asymmetry between global warming science and what the public ultimately hears. It's a good bet that, had Wöppelmann et al. found that sea level was rising at a rate 70 percent *higher* than the IPCC estimated, their findings would be on the front page of every newspaper in the world.

Greenland's ice sheets and glaciers make up the largest ice mass in the Northern Hemisphere, some 2.85×10^6 cubic kilometers (6.8×10^5 cubic miles), or 9.9 percent of total global ice volume. Together, Greenland and Antarctica hold 99.4 percent of the world's ice. The remaining nonpolar ice volume, including the vast Himalayan Ice Cap, is a mere 0.6 percent.

A 2006 *Science* paper by Eric Rignot and Pannir Kanagaratnam received a tremendous amount of publicity when it claimed that there has been a widespread and accelerating loss of Greenland's peripheral glaciers during the past 10 years, and increasing runoff from the main ice sheet, as measured by satellites. The rate given was 224 ± 41 cubic kilometers (53 ± 10 cubic miles) per year for 2005. For comparative purposes, the Greenland ice mass given above, in standard numerical notation, is 2,850,000 cubic kilometers (685,000 cubic miles), yielding a loss of eight-thousandths of a percent per year. That translates into a sea-level rise of two-hundredths of an inch per year.

Amazingly, there was no reference in this paper to Ola Johannessen's 2005 paper, in the same journal, that showed that the Greenland ice cap is *accumulating* at a rate of 5.4 ± 0.2 centimeters per year (2.1 ± 0.1 inches) That increase in the elevation of the ice cap was measured by the very same satellites that Rignot and Kanagaratnam used!

What's the difference? Rignot and Kanagaratnam combined observations of ice loss from the coastal glaciers with models of changes over the inland ice cap, whereas Johannessen et al. observed changes in the ice cap directly. Johannessen et al. found that the rise in ice-cap elevation converts to about 75 cubic kilometers (18 cubic miles) per year. Had Rignot and Kanagaratnam used *real data* as opposed to a *computer simulation*, they would have found that any loss of Greenland ice had occurred only in the last five years (it was gaining ice before then, even after accounting for the loss from the glaciers), and the total loss would be around 93 cubic kilometers (22.3 cubic miles), which is slightly more than 40 percent of the already tiny loss they originally found.

Figure 4.2 displays the temperature history for southern Greenland from the Danish Meteorological Institute, from 1782 through 2007 (understandably, some years in the late 18th and early 19th century don't have enough data). That is the area with the greatest glacial retreat. Note that temperatures from 1925 through roughly 1960 were generally higher than they are today.

Writing about the mass balance of Greenland ice in *Science* in 2000, Krabill et al. said:

> Greenland temperature records from 1900–1995 [note: Figure 4.2 is through 2007] show the highest summer temperatures

Figure 4.2
SOUTHERN GREENLAND TEMPERATURES, 1782–2007

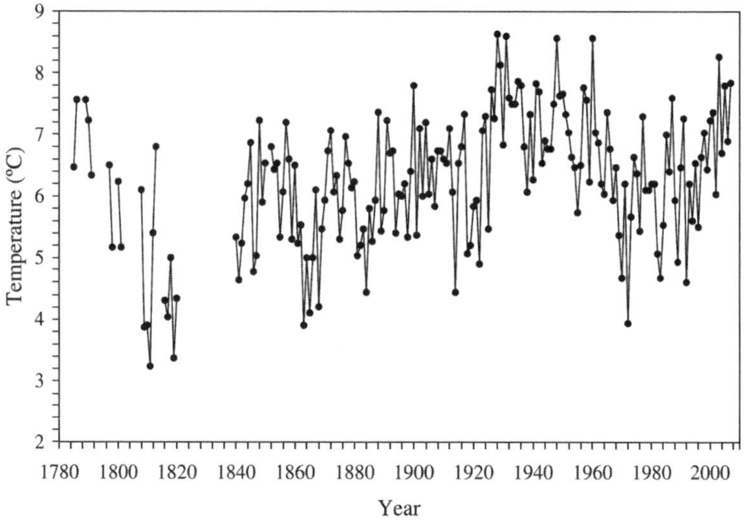

SOURCE: Danish Meteorological Institute 2008: http://www.dmi.dk/dmi/tr08-40.pdf.

> in the 1930s, followed by a steady decline until the early 1970s and a slow increase since. The 1980s and 1990s were about half a degree colder than the 96-year mean. Consequently, if present-day thinning is attributable to warmer temperatures, thinning must have been even higher earlier this century.

In 2006, Petr Chylek, from the Los Alamos National Laboratory, and colleagues wrote:

> Since 1940, however, the Greenland coastal stations have undergone predominantly a cooling trend. At the summit of the Greenland ice sheet, the summer average temperature has decreased at the rate of 2.2°C per decade since the beginning of measurements in 1987. This suggests that the Greenland ice sheet and coastal regions are not following the current global warming trend.

In 2006, Chylek et al. also put recent Greenland temperatures in perspective, particularly in the summer, when ice melts:

> 1. The years 1995 to 2005 have been characterized by generally increasing temperatures at the Greenland coastal stations. The year 2003 was extremely warm on the southeastern coast of Greenland. The average annual temperature and the average summer temperature for 2003 at Ammassalik (southeast coast) was a record high since 1895. The years 2004 and 2005 were closer to normal being well below temperatures reached in 1930s and 1940s. Although the annual average temperatures and the average summer temperatures at Godthab Nuuk, representing the southwestern coast, were also increasing during the 1995–2005 period, they generally stayed below values typical for the 1920–1940 period.
>
> 2. The 1955 to 2005 averages of the summer temperatures and the temperatures of the warmest month at both Godthab Nuuk and Ammassalik are significantly lower than the corresponding averages for the previous 50 years (1905–1955). The summers at both the southwestern and southeastern coast of Greenland were significantly colder within the 1955–2005 period compared to the 1905–1955 years.
>
> 3. Although the last decade of 1995–2005 was relatively warm, almost all decades within 1915–1965 were even warmer at both the southwestern (Godthab Nuuk) and southeastern (Ammassalik) coasts of Greenland.
>
> 4. The Greenland warming of the 1995–2005 period is similar to the warming of 1920–1930 although the rate of temperature increase was about 50 percent higher during the 1920–1930 warming period.

In 2007, Chylek et al. published another paper in *Journal of Geophysical Research* in which they developed a computer model relating the loss (or gain) in Greenland ice as a function of temperature.

Chylek et al.'s previous paper and the southern Greenland temperature history gave us a hint of what's to come. They use a concept of what they call "melt-days" that reflect the integrated warming of a particular Greenland summer. This particular study was for Western Greenland, where there are two very good long-running weather stations.

Chylek et al. concluded:

> We infer that the melt-day area of the western part of the ice sheet doubled between the mid-1990s and mid-2000s, and that the largest ice sheet surface melting probably occurred

> between the 1920s and the 1930s, concurrent with the warming in that period.

They then go on to quote Hans Ahlmann (see below) who, in 1948, noted a large loss of ice from Greenland, concurrent with the end of the warm period shown in Figure 4.2. Speaking to Hansen's hypothesis of rapid ice loss, Chylek et al. said:

> An important historical fact is that this decades-long Greenland warming apparently did not exceed a threshold for rapid ice sheet disintegration as evidenced by ice sheet stabilization and regrowth that followed.

Helheim Glacier: A Cautionary Tale

In December 2005, the BBC reported "Greenland Glacier Races to Ocean," describing the behavior of two major glaciers in eastern Greenland. The two they describe, Kangerdlugssuaq and Helheim, were in rapid retreat, with the termini receding more than two miles per year. Together, these two massive glaciers comprise about 8 percent of total drainage area of the very large island. They are the two glaciers with the largest annual "discharge" into the ocean. Gordon Hamilton, of the Climate Change Institute at the University of Maine, was quoted as saying that those movements "suggest that the predictions for both the rate and timing for sea-level rise in the next few decades will be largely underestimated."

Wouldn't you know, that during the 18 months after the BBC ran its story, both glaciers slowed down, stopped receding (despite warm temperatures), and began *advancing*?

Ian Howat, from the University of Washington, and colleagues noted in *Science* in spring 2007 things had changed dramatically: "Average thinning over the [Kangerdlugssuaq] glacier during the summer of 2006 declined to near zero, with some apparent thickening." Helheim also "decelerated."

"Decelerated"? Not really. How about "advanced"? Figure 4.3 (see insert) is a Landsat image of Helheim from August 30, 2006 (red line). You can see that the glacier advanced substantially in the last year (the black line is August 29, 2005). In fact, it has returned to beyond its position in 1933, which we found in the U.S. Geological Survey's 1995 publication "Satellite Image Atlas of Glaciers of the World: Greenland" by Anker Weidick.

Figure 4.4
SUMMER TEMPERATURES AT ANGMAGSSALIK, GREENLAND

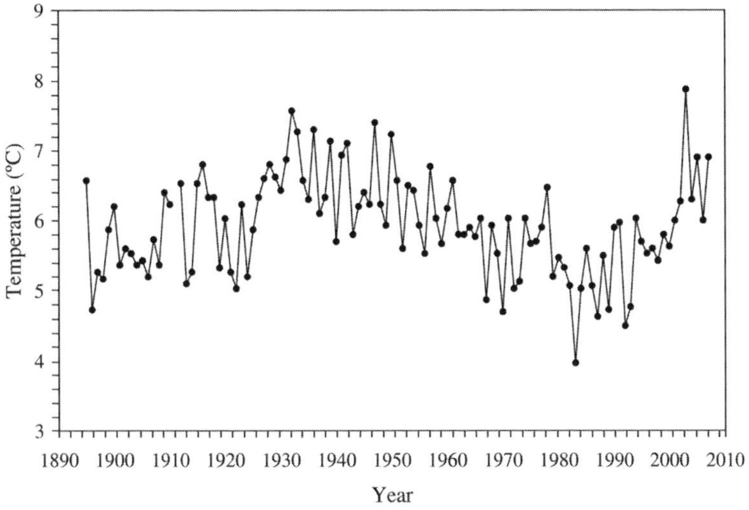

SOURCE: Danish Meteorological Institute 2008. http://www.dmi.dk/dmi/tr08-04.pdf.

It is very apparent that the one-year advance between 2005 and 2006 made up half the entire loss from May 2001 through August 2005.

We searched the BBC's website to see if it covered any of this and came up empty, even as Howat and his colleagues cautioned that

> The highly variable dynamics of outlet glaciers suggests that special care must be taken . . . particularly when extrapolating into the future, because short-term spikes could yield erroneous long-term trends.

We've also included summer temperatures at Angmagssalik (Figure 4.4), which Google Earth tells us is conveniently located only 52 miles away. It is pretty apparent that there were at least two warm decades through 1950, so Helheim was likely to have retreated far beyond its 1933 position before advancing during the cooling that extended from roughly 1950 through 1995.

A bit farther to the north of Helheim and Kangerdlugssuaq glaciers, Britannia glacier—carefully mapped out in the early 1950s by a Great Britain expedition—is shown in recent satellite photographs to currently be *larger* and farther-reaching that when it was first visited (Figure 4.5; see insert).

More Scary Greenland Stories

In its "Fourth Assessment Report" on climate change, the IPCC summarized a large number of climate models for Greenland in the 21st century. On average, the IPCC projected a rise in sea level of 2 inches or so as a result of a net loss of ice from Greenland.

Yet we are bombarded by stories that Greenland is shedding ice at a tremendous rate, and that even a small amount of additional warming will result in a massive instability that will crash much of its ice into the sea by 2100, raising sea level nearly 20 feet.

In *very* large type, the *New York Times* of January 16, 2007, proclaimed "The Warming of Greenland." Rather than consulting the latest in the refereed scientific literature, the *Times* relied on an off-the-cuff estimate of ice loss given to them by Professor Carl Boggild from the University Center at Svalbard, an archipelago about halfway between Norway and the North Pole. According to the *Times*, Boggild "said Greenland could be losing more than 80 cubic miles of ice per year." The real amount determined by meticulous analysis of recent satellite data is around 25 cubic miles per year, published by NASA's Scott Luthcke in *Science* two months before the *Times* piece. (Note that that number is about half the rate claimed in the earlier Rignot study.) Twenty-five cubic miles per year is the same mean value estimated by Andrew Shepard and Duncan Wingham in a summary of recent literature that they published in *Science* in 2007.

Rather than citing a mainstream estimate, such as the IPCC's, the *Times* quoted Richard Alley, from Penn State, who stated that "a sea-level rise of a foot or two in the coming decades is entirely possible."

What does "coming decades" mean? We should expect a bit more precision from scientists. Does Alley mean the *next* decade, which Al Gore alluded to at the beginning of in the *Larry King Live* interview we quote from at the beginning of this book? Or does it mean "some time in this century"? How *many* decades?

The current sea-level rise contributed by the Greenland ice loss is too small to even be able to measure in the next decade or two. The satellite data show a reduction of four hundred-thousandths of

Greenland's total ice per year (whereas Boggild's figure "could" be around twelve hundred-thousandths). Multiplying the satellite-based fraction by the 23 feet of sea-level rise that would result if all of the ice were lost results in a current Greenland-induced rise in sea level of 0.01 inch per year. Averaged over three decades, that's a third of an inch, which is indeed too small to be detectable. Over a century, the rise becomes a bit more than an inch. Boggild's unpublished "guesstimate" yields 3.5 inches per century.

In fact, there's nothing very new going on in Greenland. Although the *Times* paid great attention to ice-loss in eastern Greenland caused by recent temperatures, it conveniently forgot to look at nearby temperature histories (as well as the overall one shown as Figure 4.2). The longest record in that region is from Angmagssalik (Figure 4.4). In the summer (when Greenland's ice melts) the temperature has averaged 6.1°C (43.1°F), over the last 10 summers, which is very close to the average for the entire record. There's one very warm summer, in 2003. The other nine years aren't unusual at all.

From 1930 through 1960, the average was 6.5°C (43.7°F). In other words, it was warmer *for three decades*, and there was clearly no acceleration in sea-level rise. What happened between 1945 and the mid-1990s was *a cooling trend*, with the period 1985–95 being the coldest in the entire Angmagssalik record, which goes back to the late 19th century. Only in recent years have temperatures begun to look like those that were characteristic of the early 20th century.

The *Times* could have written pretty much the exact same story in 1948, before humans had much of a hand in global warming. That's when Hans Ahlmann wrote, in the *Geographical Journal*, a publication of the British Royal Geographic Society: "The last decades have reduced the ice in some parts of Greenland to such an extent that the whole landscape has changed in character." So it's hardly something new when the *Times* reports, almost 60 years later, that temperatures in Greenland "are changing the very geography of coastlines."

Ahlmann prepared a booklet accompanying a lecture to the American Geographical Society in the fall of 1953 on "Glacier Variations and Climatic Fluctuations," describing receding glaciers and rising temperatures across many disparate regions of the Arctic, as well as changes in plant life and animal behavior, and range shifts that have accompanied the climate warming. The northern migration of codfish in the Atlantic brought the species into southern Greenland for the first time in recent memory, and ushered newfound prosperity into the region. Ahlmann quotes the Prime Minister of Denmark:

> In the last generation changes that have had a decisive influence on all social life have occurred in Greenland. A new era has begun. These changes are primarily due to two circumstances. Firstly, the Greenland climate has changed, and with it Greenland's natural and economic prospects.

The "Discovery" of "Warming Island"

> A peninsula long thought to be a part of Greenland's mainland turned out to be an island when a glacier retreated. . . . [The] ominous implications are not lost on [Dennis] Schmitt, who says he hopes that the island *he discovered* [italics added] in Greenland in September will become an international symbol of the effects of climate change. Mr. Schmitt, who speaks Inuit, has provisionally named it Uunartoq Qeqertoq: The Warming Island.
>
> —John Collins Rudolf, *New York Times,* January 16, 2007

"Warming Island" was such a hit with the environmental community that it generated its own website, http://www.warmingisland.org. Rudolf bragged about his news exploits on his blog:

> I wrote a story about the new island for *The New York Times.* . . . A short video posted on the Internet [appeared] on ABC, BBC, CBC . . . in May 2007 Dennis Schmitt returned to Warming Island with Anderson Cooper of CNN for a live broadcast about climate change.

"Warming Island" is a pretty distinctive place, with a very odd shape, comprising three long "fingers" (Figure 4.6; see insert). The figure shows the loss of the ice bridge between it and the mainland, based upon satellite imagery from 1985, 2002, and 2005. When the loss of ice revealed open water, it became apparent that this land was in fact an island.

Another image (Figure 4.7; see insert), taken from land, reveals the obvious separation between Warming Island and the mainland.

Note the general surroundings of eastern Greenland in the area map (Figure 4.8; see insert). To the left of the area of concern is "Carlsbad Fjord."

The history of Southern Greenland's temperature clearly reveals a much larger integrated warming in the early and mid-20th century than the current decade. This prompts an obvious inquiry. If 10 years of not-so-unusual temperatures revealed that "Warming Island" was indeed an island and not a peninsula, was it also an island at the end of the last warm period, roughly 50 years ago?

In the early 1950s, Switzerland's Ernst Hofer spent four summers as an aerial photographer in support of ground-based geological research and mapping efforts. In 1957, he published a remarkable book about northeast Greenland, entitled *Arctic Riviera* (Figure 4.9; see insert; our copy of this rare book was quite damaged, so the cover isn't reproduced in its entirety).

Remember that 1957 was near the end of several decades that averaged warmer than the most recent 10 years. In the introduction, Danish explorer Lauge Koch praised the regional climate:

> [Hofer] has indeed given a characteristic description of the fjord-region of North-East Greenland, which, owing to favorable circumstances, enjoys a distinctly mild climate. . . . During this [summer] period the glaciers supply enough water to produce a small Arctic oasis . . . the midnight sun warms the steep walls of the fjords and produces temperatures that can otherwise rarely be registered in such northern degrees of latitude.

Arctic Riviera includes a map of Northeastern Greenland, shown in Figure 4.10 (see insert). Warming Island is shown as an island!

What's remarkable about the Warming Island story is that every scientist who has researched Greenland temperatures knows of the warmth of the early 20th century, and yet no one rose to question the claims that "for unknown centuries" it had been assumed to be part of the mainland.

NASA: Greenland Bigger than U.S.!

From a September 25, 2007, NASA press release:

> A new NASA-supported study reports that 2007 marked an overall rise in the melting trend over the entire Greenland ice sheet and, remarkably, melting in high-altitude areas was greater than ever at 150 percent more than average. In fact, the amount of snow that has melted this year over Greenland could cover the surface size of the U.S. more than twice.

A record high in high places? High-altitude melting "greater than ever"? Amount of snow melt this year "could cover the surface size of the U.S. more than twice"? At what depth?

The NASA press release also included the following graphic, which, in these writers' humble opinions, takes the ice-cream cake for rhetorical chutzpah in the field of scientific data presentation (Figure 4.11).

Figure 4.11 shows the annual "melt area index" of Greenland in relation to the size of the United States for each year from 1988 to 2007. The value for this year is a bit more than two times the size of the continental United States. Now, considering that the *total* area of Greenland is just more than one-quarter the area of the lower 48, you may wonder how an area of more than *twice* the size of the continental United States melted this year in Greenland.

The answer lies in what exactly the "melt area index" represents. Readers probably think it is the area of Greenland where there is some snowmelt over the year.

In this case, the "United States" units represent in fact the *sum* of the area of Greenland that experienced surface snowmelt across all days of the year that melting occurred. That is to say, if an area of Greenland equal to 1/365 the area of the United States experienced melting every day of the year, that would produce a "melt area index" for that location equal to the size of the entire contiguous United States.

Rather than using a picture of the continental United States as a metric in its graph, even though it would have been less sensational, NASA should simply have plotted out the time history of the "melt area index" for Greenland and left it at that (we've done this service for you in Figure 4.12).

As is obvious, there is a general rise since NASA satellite records began in 1988, *but it is all confined to the period 1988–97*. Since then, the "melt area index" varies from year to year, but there is no overall net change.

What's the news here? Six of the last ten years have a larger "melt area index" than 2007. If, instead of calculating the "melt area index" for all of Greenland, you limited the calculation to only those regions that lie at elevations above 2,000 meters (1.25 miles, or, in NASA parlance, "high places"), then 2007 is indeed the highest on record since 1988 (or, as NASA described it, "greater than ever") (see Figure 4.13).

Figure 4.11
MELTING INDEX TREND IN GREENLAND, 1988–2007, IN RELATION TO U.S. SURFACE AREA

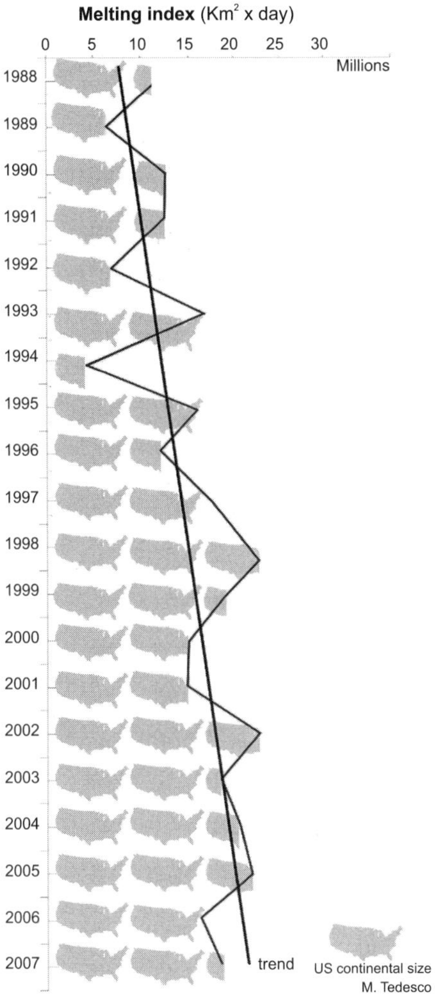

SOURCE: NASA 2007. http://www.nasa.gov/vision/earth/environment/greenland_recordhigh.html.

NOTE: km^2 = square kilometer.

Figure 4.12
GREENLAND'S MELT AREA INDEX, 1988–2007

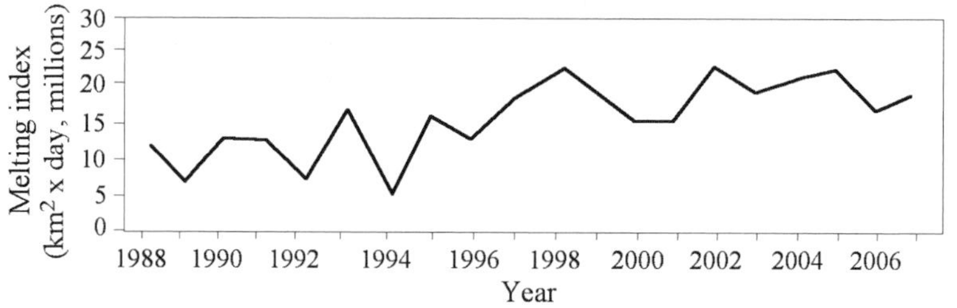

SOURCE: Adapted from NASA September 25, 2007, Press Release.
NOTE: km^2 = square kilometer.

Figure 4.13
MELT AREA INDEX OF THE GREENLAND ICE SHEET ABOVE 2,000 METERS, 1988–2007

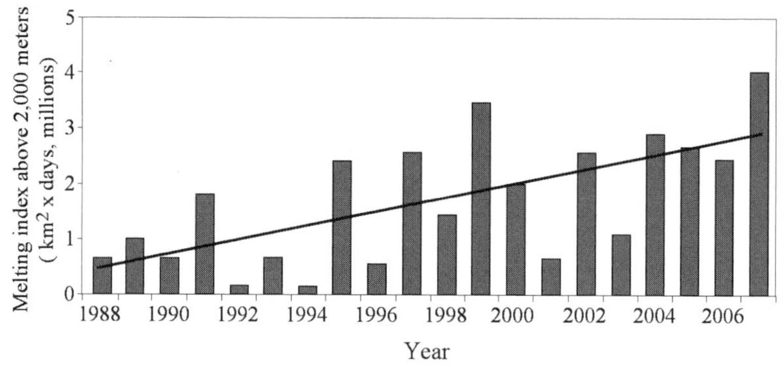

SOURCE: Adapted from Tedesco 2007.
NOTE: 2,000 meters = 1.25 miles; km^2 = square kilometer.

Inquiring minds might want to know what the high-elevation melt area was during the warm period in the early and mid-20th century. Obviously, we didn't have satellites taking measurements from space back then, but there was a good deal of climate research taking place on the ground across Greenland. In 1961, much of this work was summarized in an article by R. W. Gerdel titled, "A Climatological Study of the Greenland Ice Sheet." This was a part

Figure 1.5

PROJECTED WARMING TRENDS BASED ON COMPUTER MODELS FOR THE MIDRANGE SCENARIO FOR CARBON DIOXIDE EMISSIONS, 2000–2100

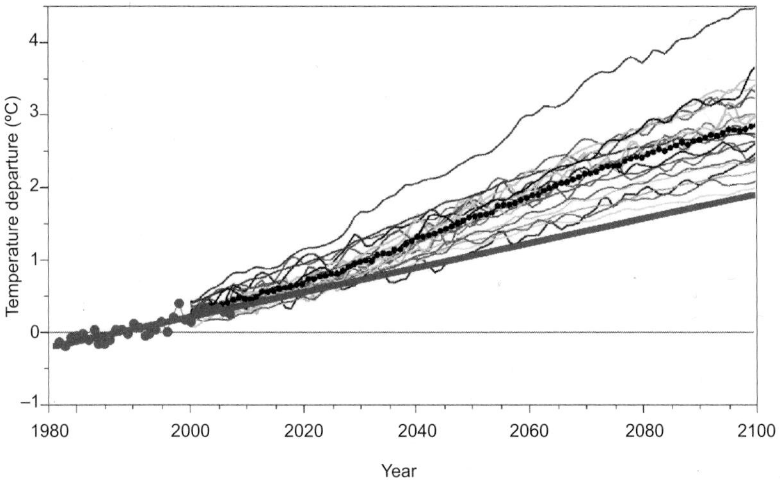

SOURCE: IPCC, 2007.

NOTE: The colored lines represent projected warming trends based on various climate models. The black dots represent the projected trends' average for the IPPC's midrange scenario for carbon dioxide emissions. The red line supperimposes observed temperatures from 1975 through 2007, and a projection of that rate through the end of the century.

Figure 2.13
DIFFERENCES BETWEEN OBSERVED AND ADJUSTED TRENDS AROUND THE WORLD (UNITS = °C/DECADE)

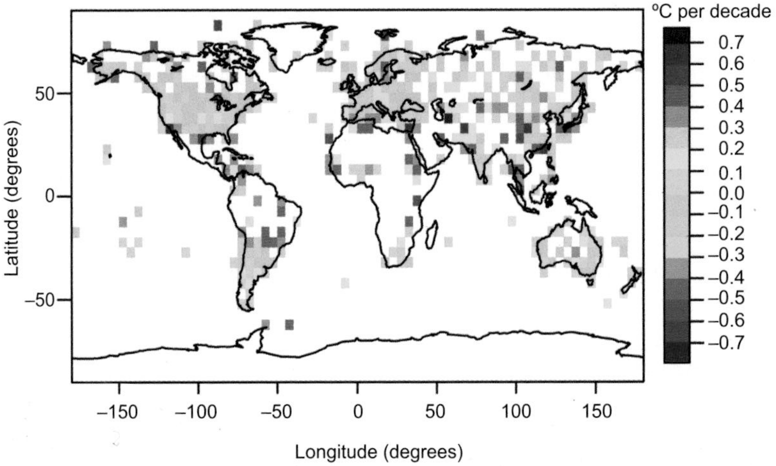

SOURCE: McKitrick and Michaels, 2007.

Figure 3.3

COMPARISON OF HURRICANE SEASON MAPS OF 1933 (BOTTOM RIGHT) AND 2005 (TOP LEFT)

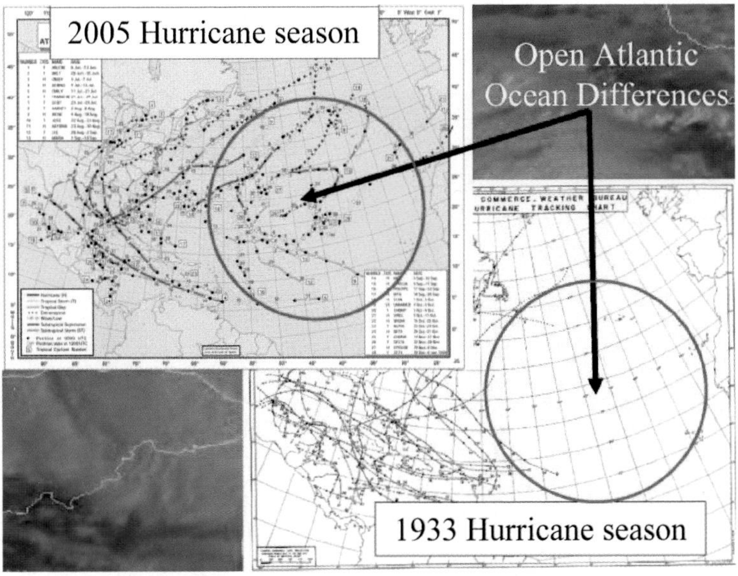

SOURCE: Landsea, 2007.

Figure 4.3

Helheim Glacier, Central Eastern Greenland Coast, 1933–2006

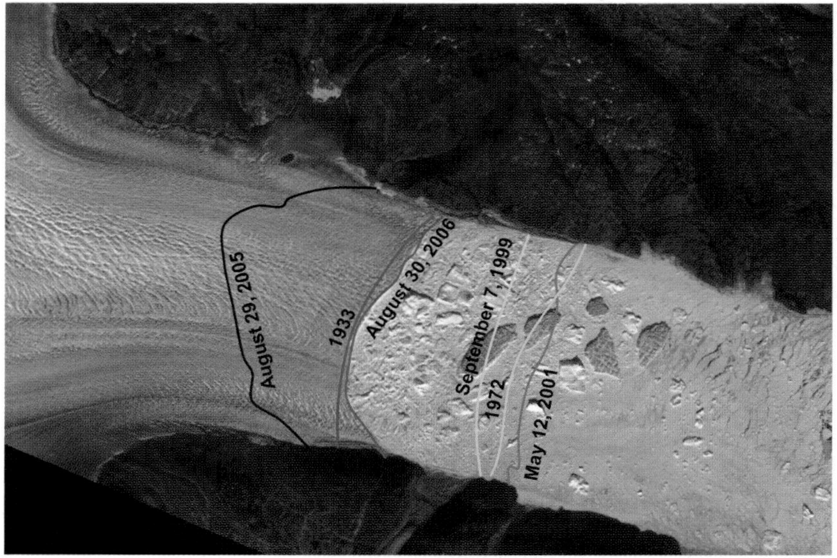

Sources: Landsat Imagery; Howat et al., 2007; Weidick, 1995.

Note: From September 1999 (peach line) to May 2001 (orange line) the Helheim Glacier advanced slightly, pushing its calving front beyond its location in 1972 (green line). A slow retreat that began in 2001 was followed by a rapid retreat from 2004 to August 2005 (black line). Thereafter, the glacier stopped receding and began advancing again. By August 2006, the calving front had advanced beyond its location in 1933 (blue line) and is again approaching its summer 2004 location.

Figure 4.5

BRITANNIA GLACIER, GREENLAND, 2008 POSITION (A) AND 1954 POSITION (B)

a. 2008 b. 1954

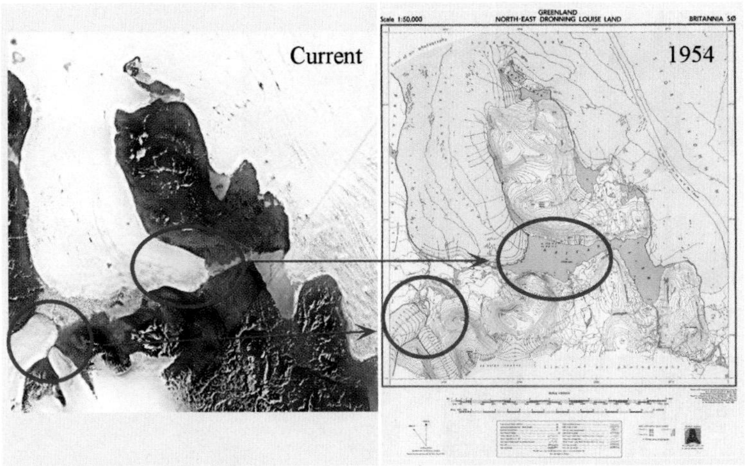

SOURCES: Yahoo! Maps, 2008; Hamilton et al., 1956.

NOTE: Figure 4.5a shows the current position of the Britannia Glacier as captured from a satellite photo available from Yahoo! Maps. Figure 4.5b is a detailed map of the position of the same glacier produced from photographs and a ground survey done in 1954 (Hamilton et al., 2006). Currently the Britannia Glacier and a smaller side glacier are advanced beyond their 1954 termini (red circles).

Figure 4.6
WARMING ISLAND, GREENLAND, 1985, 2002, AND 2005

a. August 11, 1985 b. September 5, 2002 c. September 4, 2005

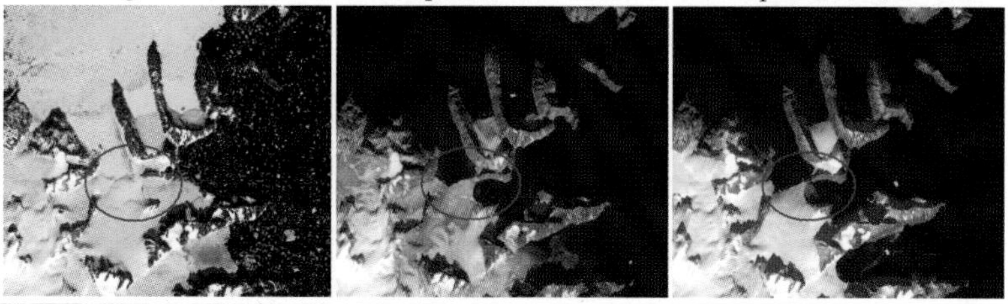

SOURCE: U.S. Geological Survey, 2005.

Figure 4.7
WARMING ISLAND, GREENLAND, 2006

a. Overhead View b. Oblique View

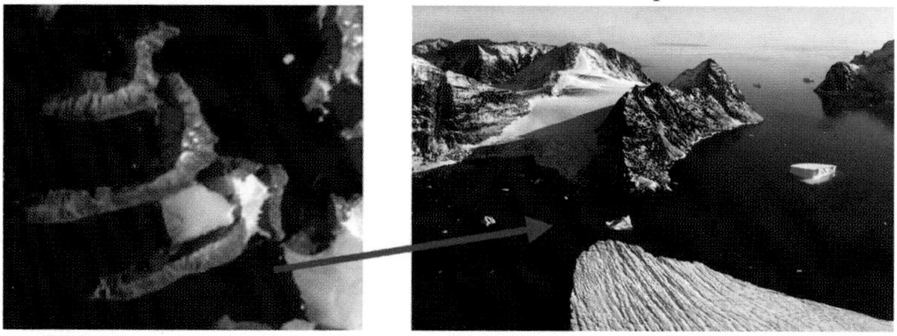

SOURCE: Adapted from U.S. Geological Survey, 2005; *New York Times*, January 16, 2007.

Figure 4.8
MAP OF WARMING ISLAND, GREENLAND

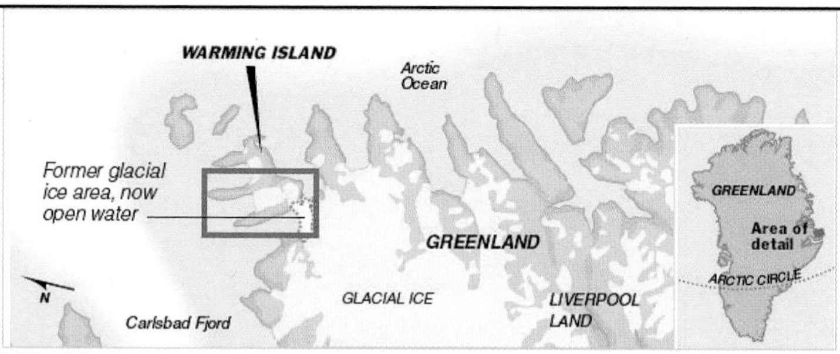

SOURCE: *New York Times*, January 16, 2007.

Figure 4.9
COVER OF *ARCTIC RIVIERA*

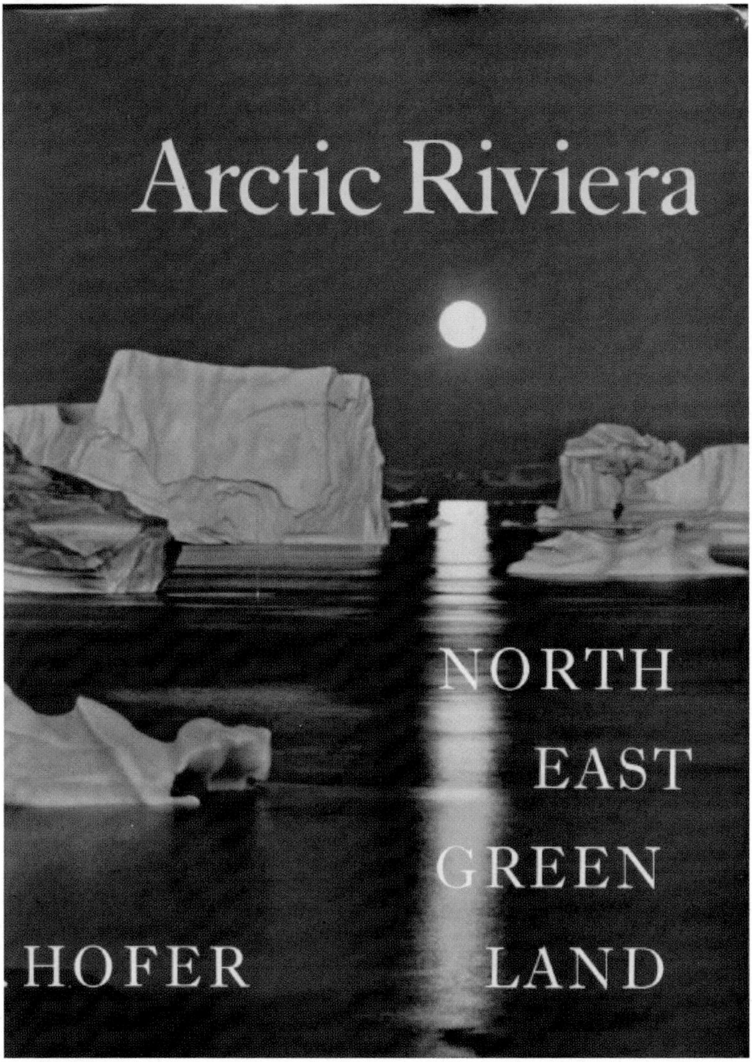

SOURCE: Hofer, 1957.

Figure 4.10
WARMING ISLAND, GREENLAND, 1957

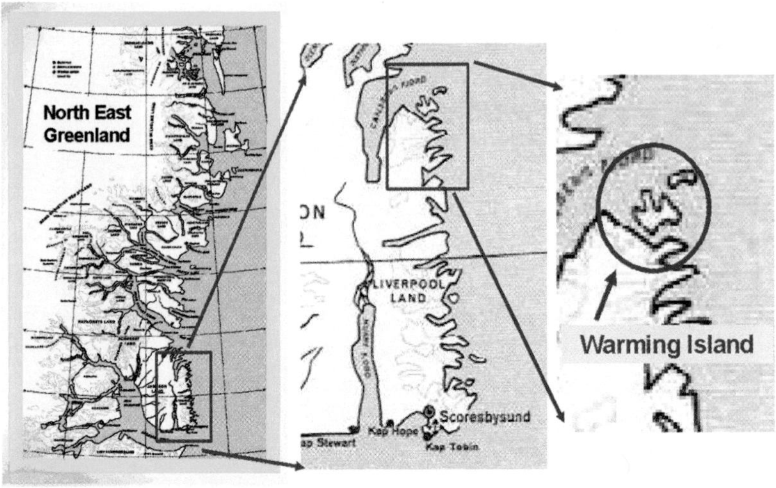

SOURCE: Hofer, 1957.

Figure 4.14
DAYS WITH MELTING IN 2006; OBSERVED (COLORS) AND CALCULATED IN 1961 (NUMBERS)

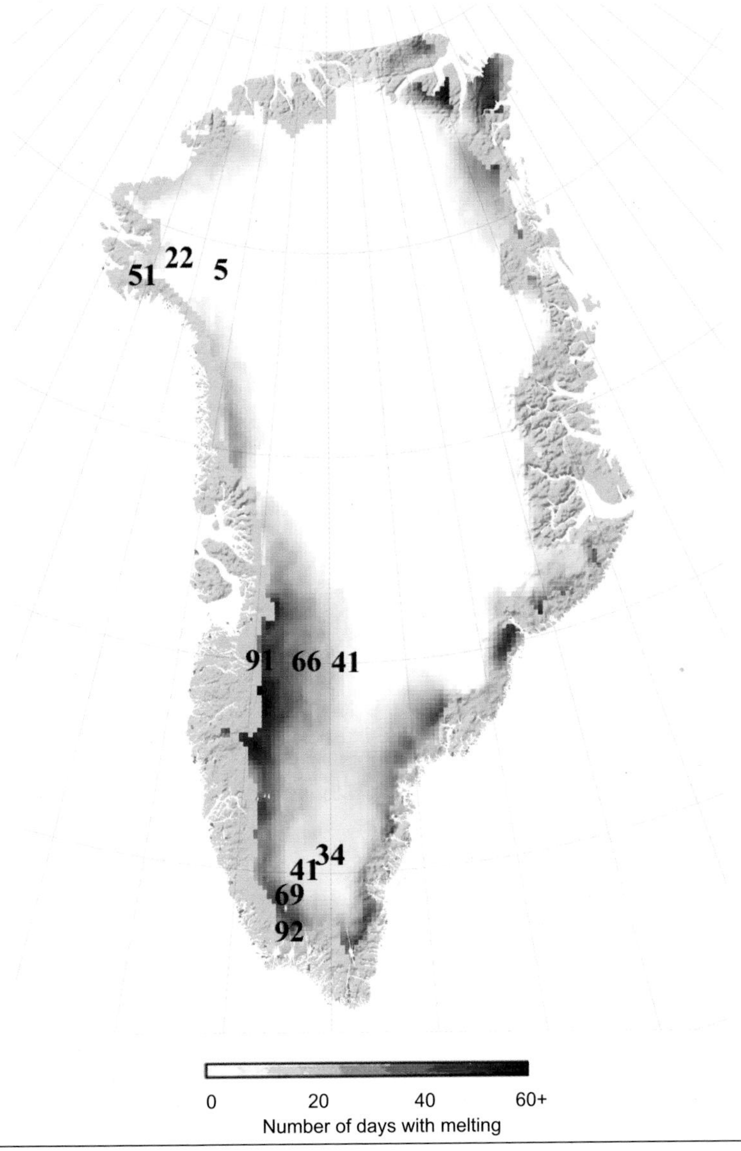

SOURCES: NASA, 2007; Gerdel, 1961.

Figure 4.19

TEMPERATURES RELATIVE TO PREINDUSTRIAL LEVELS WORLDWIDE, LAST 12,000 YEARS, FIRST DRAFT OF THE IPCC'S 2007 REPORT

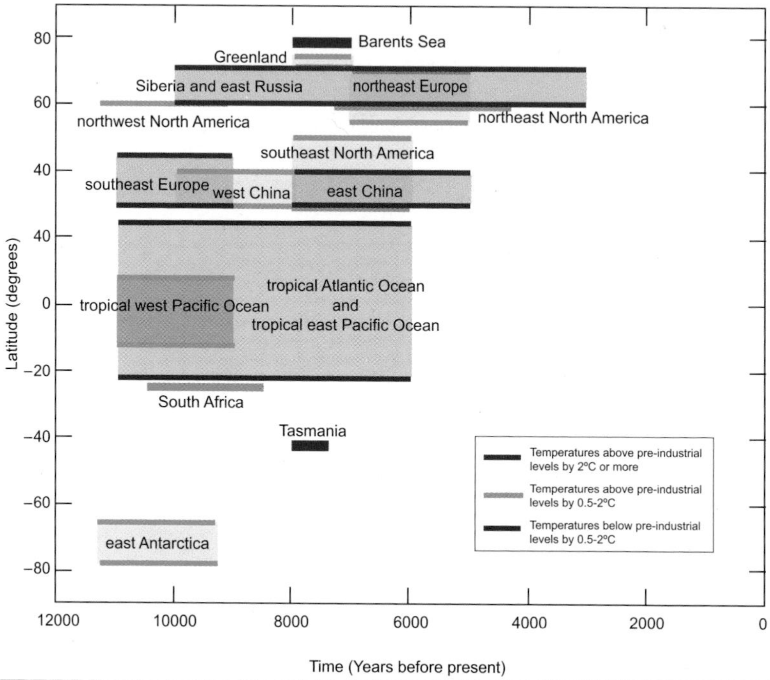

SOURCE: First Order Draft, Fourth Assessment Report, IPCC, 2007.

Figure 4.20

TEMPERATURES RELATIVE TO PREINDUSTRIAL LEVELS WORLDWIDE, LAST 12,000 YEARS, SECOND DRAFT OF THE IPCC'S 2007 REPORT

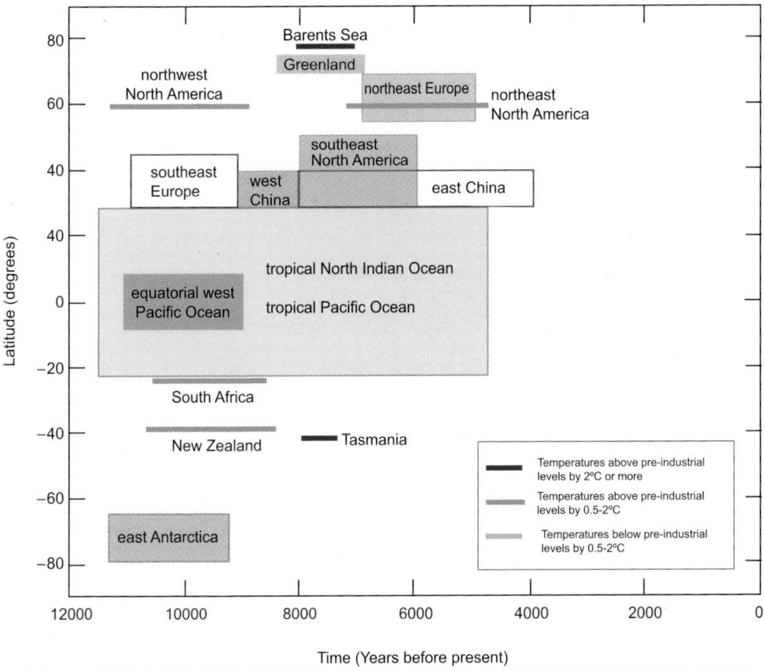

SOURCE: Second Order Draft, Fourth Assessment Report, IPCC, 2007.

Figure 4.21
TEMPERATURES RELATIVE TO PREINDUSTRIAL LEVELS WORLDWIDE, LAST 12,000 YEARS, PUBLISHED VERSION OF THE IPCC'S 2007 REPORT

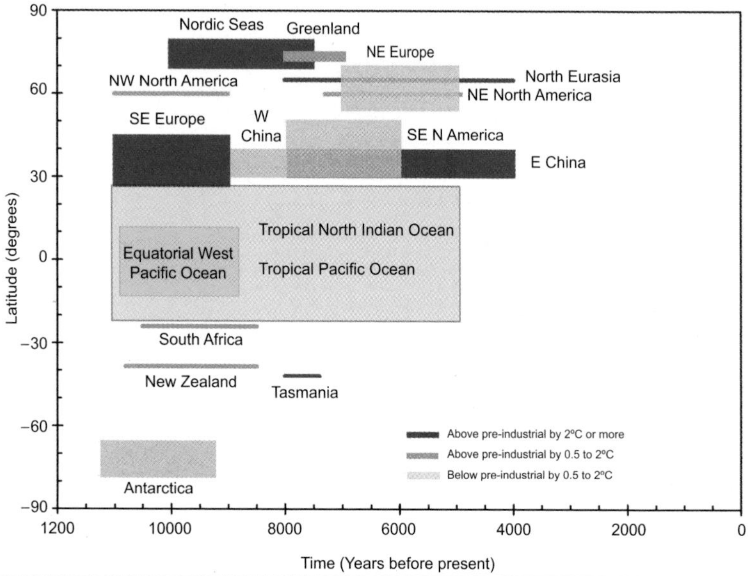

SOURCE: IPCC, 2007.

Figure 4.22
NEWTOK, ALASKA, 2007

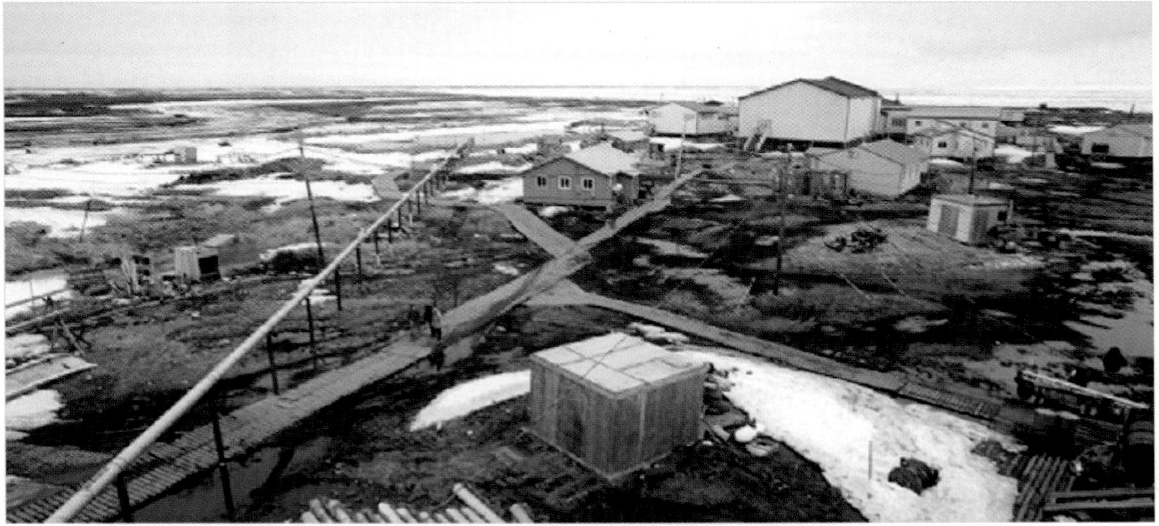

SOURCE: *New York Times*, May 27, 2007.

NOTE: The photograph's caption read, "Newtok, Alaska, in spring, as viewed from its water tower. Boardwalks squish into the muck in Newtok, which erosion has turned into an island."

Figure 4.23

Linear Trends of Annual Mean Surface Air Temperature, 1958–2002

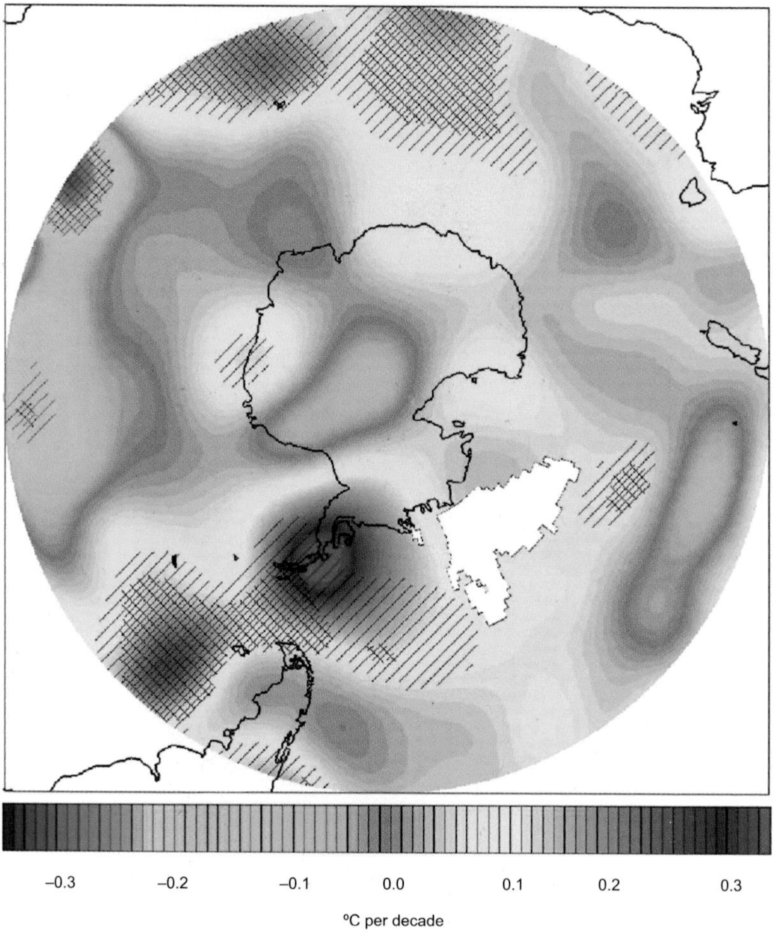

Source: Chapman and Walsh, 2007.

Figure 6.1
GLOBAL THICKNESS ANOMALIES, JUNE, JULY, AND AUGUST 2003

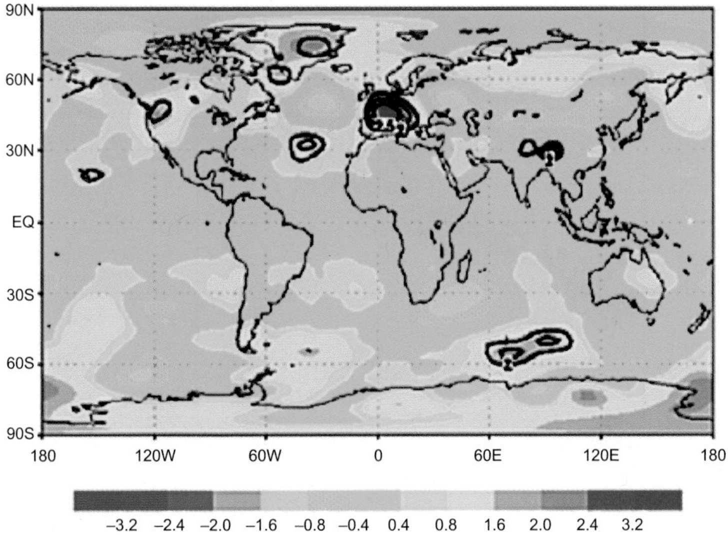

SOURCE: Chase et al., 2006.

NOTE: These anomalies are proportional to the average temperature of the bottom half of the atmosphere. Areas exceeding 2.0, 2.5, and 3.0 standard deviations from the 1979–2003 mean are contoured in thick lines for anomalies of both signs.

Figure 7.1
RESULTS OF FIRST "COUPLED MODEL INTERCOMPARISON PROJECT"

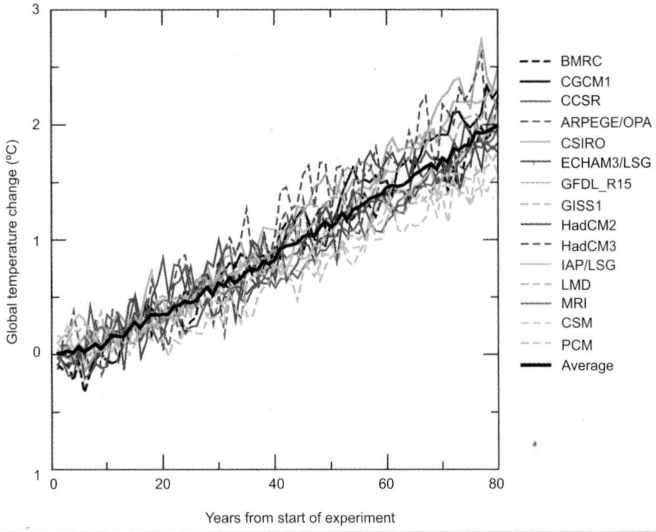

SOURCE: Meehl et al., 2001.
NOTE: Acronyms refer to various climate models.

Figure 7.3
THE "HOCKEY STICK" AS IT APPEARED IN THE IPCC'S THIRD ASSESSMENT REPORT

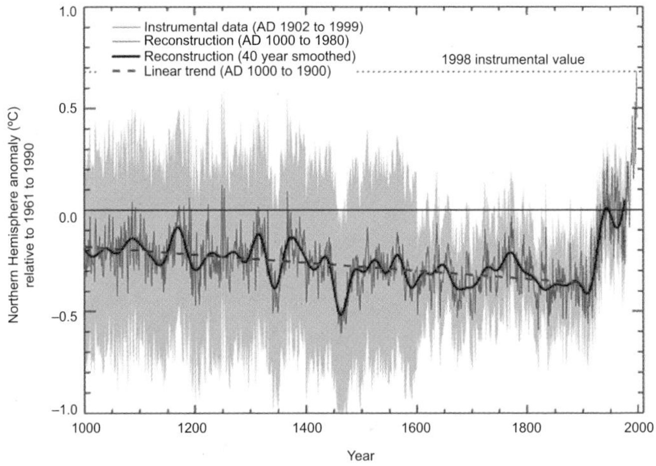

SOURCE: IPCC, 2001.
NOTE: Michael Mann's original *Nature* paper featured a 600-year history.

Table 4.1

AVERAGE NUMBER OF SUMMER DAYS WITH MAXIMUM TEMPERATURE CALCULATED TO BE AT OR ABOVE FREEZING FOR VARIOUS ELEVATIONS AND LATITUDES ACROSS GREENLAND

	Elevation (meters)								
Latitude (°N)	*500*	*1,000*	*1,300*	*1,500*	*1,700*	*2,000*	*2,500*	*2,700*	*3,000*
76	73	51	32	22	14	5	0	0	0
67	91	91	89	86	81	66	41	34	16
61	92	92	91	90	86	69	41	33	15

SOURCE: Gerdel 1961.

of the 1961 *Symposium on the Physical Geography of Greenland of the XIX International Geographical Congress*. Included among Gerdel's discussions of temperature, precipitation, winds, fog, radiation, and so on is a section called "The Occurrence of Melting on the High Ice Sheet."

Gerdel reports that there is evidence of summer melting occurring at least as high as 1,700 meters (1 mile) above sea level on the interior of the ice sheet east of Thule, at latitude 76°N. That is extremely far north, less than a thousand miles from the North Pole. Air temperature measurements from the Thule air base (along the northwestern Greenland coast) coupled with those taken from elevations on the ice sheet indicate that the temperature "lapse rate" (decline in temperature with height) were found to be 0.6°C (1.0°F) per 100 meters. Gerdel used that lapse rate to calculate the elevation on the ice sheet where the air temperature would reach the freezing point, extrapolated from the temperatures taken from coastal stations around Greenland (there were and are very few temperature measurements from locations on the interior ice sheet itself). From those data, Gerdel produced Table 4.1, which indicates the calculated average number of days during the summer (June, July, and August) that the maximum air temperature on the ice sheet was at or above freezing for various latitudes and elevations. This is based upon temperatures observed during the period 1946–56.

As Table 4.1 shows, there's likely to have been plenty of melting at altitudes as high as 2,700 meters (1.7 miles).

As our Figure 4.12 illustrates, the "melt area index" of 2006 was very close to the 2007 value. A comparison of Gerdel's 1961 numbers with the 2006 data provides a fair assessment of how current conditions at high elevation compare with mid-20th century ones (Figure 4.14; see insert).

Figure 4.14 (color insert) shows the number of days with melting observed across Greenland in 2006 as reported by NASA, along with the number of days of melting for the locations as calculated by Gerdel in 1961. Notice that in every case, Gerdel calculated a *greater* number of days with melting than occurred in 2006 including in the "high places" on the ice sheet in the north, south central, and southern portions of Greenland.

Arctic Sea Ice in Perspective

A September 2005 press release highlighted a decrease in satellite-sensed Arctic ice extent from September 1979 (the beginning of the record) to September 2005, with that year showing the lowest values in the entire record. Note that this is a seasonal phenomenon. The ice extent reaches its annual minimum sometime around mid-September. On the first day of autumn, night falls at the North Pole and the sun does not return for six months. The ice soon begins to re-form.

Nowhere does NASA's press release mention that 1979 was very close to the end of the second-coldest period in the Arctic in the 20th century (Figure 4.15). Because temperatures in 1979 had just recovered from their lowest values since the early 1920s, Arctic ice should have been near a maximum for the last eight decades when the first satellite imagery was returned.

The September 2005 record minimum for Arctic sea ice was broken again in September 2007, making headlines worldwide. (As noted later in this chapter, within a few months, satellites would detect record *high* ice extent in the Southern Hemisphere.)

Is this unprecedented? A very interesting 2003 study by National Oceanic and Atmospheric Administration scientists James Overland and Kevin Wood examined the logs of 44 Arctic exploration vessels from 1818 to 1910 and found that "climate indicators such as navigability, the distribution and thickness of annual sea ice, monthly surface air temperatures, and the onset of melt and freeze were within the present range of variability." Commenting on the early exploration logs, they noted that "overwinter locations of Arctic

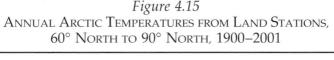

Figure 4.15
ANNUAL ARCTIC TEMPERATURES FROM LAND STATIONS, 60° NORTH TO 90° NORTH, 1900–2001

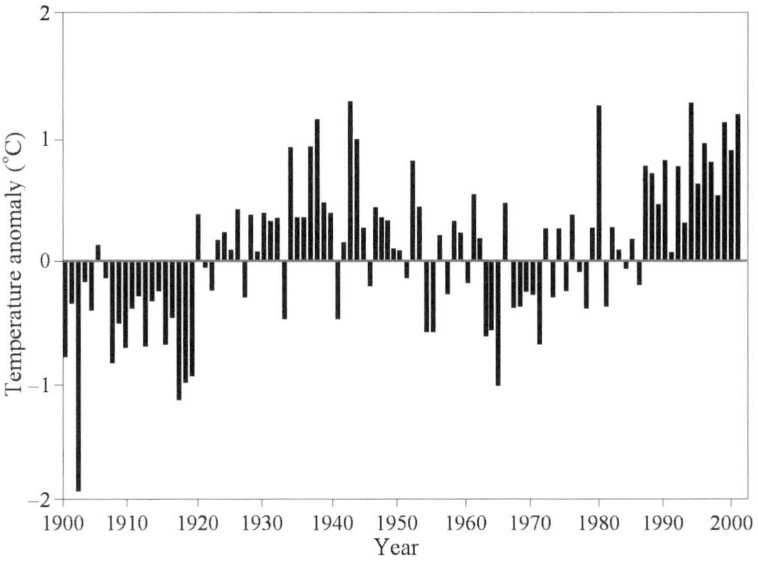

SOURCE: Adapted from *Arctic Climate Impact Assessment* 2004. http://www.acia.uaf.edu/pages/overview.html.

discovery expeditions from 1818 to 1859 are surprisingly consistent with present sea ice climatology."

It's easy to blame recent Arctic warming on greenhouse gases. After all, computer models all show that warming is enhanced in the high northern latitudes, more so than on the rest of the planet. So the Arctic should give some of the earliest signals of change.

The warming that peaked around 1940 remains troublesome, though. Given the forecast strength of the greenhouse signal in the Arctic, are present temperatures unprecedented? And what caused the large and rapid warming of the early 20th century (1900–40), whose magnitude of 2°C (3.6°F) isn't statistically distinguishable from the amount of warming in the most recent four decades?

Writing in the journal *Geophysical Research Letters* in 2003, Vladimir Semenov and Lennart Bengstsson, from Germany's Max Planck Institut, found that the recent arctic temperature rise is largely related to atmospheric circulation factors in the North Atlantic region, while the early 20th-century warming was probably because of sea ice variations.

This creates further problems concerning the current warming. One of the reasons that the Arctic is forecast to warm up more than other places is because the ice disappears. Here's why: The bright ice reflects away much of the sun's energy, whereas the darker water absorbs it. So when the ice goes down, the temperature should go up, which melts more ice, which raises the temperature more, and so on—an example of a "positive feedback loop" in the climate system. Indeed, in the 2005 NASA press release, National Snow and Ice Data Center senior scientist Ted Scambos confirmed as much, saying, "Feedbacks in the system are starting to take hold."

It's fair to say that Semenov and Bengtsson's study casts doubt on an irreversible positive feedback. If the substantial warming of the early 20th century indeed resulted from sea ice changes, then why did the warming not continue? The fact is that the cause of the substantial arctic cooling trend from 1940 through the mid-1970s remains mysterious.

Gore-ing of History

Obviously, there was a low point in Arctic ice during the last temperature maximum, observed in the 1930s. But you wouldn't get that from the book version of Al Gore's science fiction classic *An Inconvenient Truth.*

Shown below, as Figure 4.16, is a figure Gore labeled "Sea-Ice Extent: Northern Hemisphere."

Gore's depiction of the Northern Hemisphere sea ice extent (Figure 4.16) shows basically small annual variations, but no trend from about 1900 through about 1970, and then a large decrease. The loss of ice extent certainly *looks* pretty dramatic and gives the distinct impression that human activities are producing changes that are quite unusual, at least in the context of the last 100 years.

As additional historical data and analyses come to light, however, it is looking less and less likely that the early-to-mid-20th-century variations in Arctic sea ice were as small as Gore claims.

In his 1953 American Geographical Society pamphlet, Ahlmann noted the following:

Figure 4.16
ARCTIC SEA-ICE EXTENT, 1900–2005,
FROM *AN INCONVENIENT TRUTH*

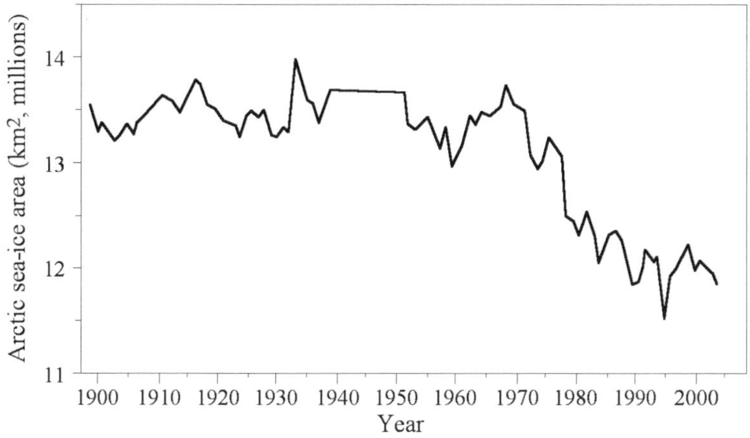

SOURCE: Adapted from Gore 2006, p. 143.
NOTE: km^2 = square kilometer.

> The thickness of the ice forming annually in the North Polar Sea has diminished from an average of 365 centimeters [12 feet] at the time of Nansen's Fram expedition of 1893–96 to 218 centimeters [7.2 feet] during the drift of the Russian icebreaker Sedov in 1937–40. The extent of drift ice in Arctic waters has also diminished considerably in the last decades. According to information received in the U.S.S.R. in 1945, the area of drift ice in the Russian sector of the Arctic was reduced by no less than 1,000,000 square kilometers [386,100 square miles] between 1924 and 1944. The shipping season in West Spitsbergen has lengthened from three months at the beginning of this century to about seven months at the beginning of the 1940s. The Northern Sea Route, the North-East Passage, could never have been put into regular usage if the ice conditions in recent years had been as difficult as they were during the first decades of this century.

In a September 21, 2007, seminar at University of Colorado, U.S. Department of Commerce scientist Andy Mahoney explained:

> [Historic] time series of air temperature and the extents of pack ice, multiyear ice and landfast ice extents reveal three

> distinct periods of variability over the last eight decades: a period of warm winters and decreasing summer and fall sea ice extent (period A), followed by a cool period of stable or slightly increasing extent (period B) before a period of year-round warm temperatures and ice loss (period C).

What's wrong with Gore's picture? Gore wants to relate recent Arctic sea-ice declines to the recent warm-up there. But since the record of Arctic temperatures shows not only a warm-up in recent decades, but also one similar in relative magnitude from the early years of the 20th century to about the mid-1940s (Figure 4.15), shouldn't there have been a loss in sea ice in both periods? Wasn't Gore suspicious that his figure didn't show one? And what about the period from the mid-1940s to the mid-1970s when Arctic average temperatures declined a healthy amount? Shouldn't there have been an increase in sea ice extent during that period?

It is not acceptable for Gore to hide behind the source of his figure, which he lists as "Hadley Carter." (Who is *that*? We'll bet it means "Hadley Center," a British government entity.) The closest thing we can find is Figure 4.17, the annual data from the Cryosphere Today website of the University of Illinois Polar Research Group (available at http://arctic.atmos.uiuc.edu/cryosphere/IMAGES/seasonal.extent.1900-2007.jpg). But Gore made one critical omission: Cryosphere Today's accompanying text explains that the data prior to 1953 are pretty unreliable.

Accompanying the Cryosphere Today figure is a data set documentation file (http://arctic.atmos.uiuc.edu/SEAICE/arctic.historical.seaice.doc.txt) that is full of caveats about how the data set was put together. It cautions: "Please note that much of the pre-1953 data is either climatology or interpolated data and the user is cautioned to use this data with care." Well, the incorporation of climatology (long-term averages) goes a long way toward explaining the lack of variation in early 20th-century data. And this is obviously what was done: note the autumn line is completely flat prior to 1953. Nowhere in the book *An Inconvenient Truth* is any of this made clear. Instead, we just see a graph with little ice variability for 70 years, and then a steep drop-off during the past 30.

And finally, a paper was published in 2004 (before *An Inconvenient Truth* was released or the book of the same title published) that discussed some Arctic ice data that wasn't included in the data set

Figure 4.17
ANNUAL SEA ICE EXTENT, 1900–2007

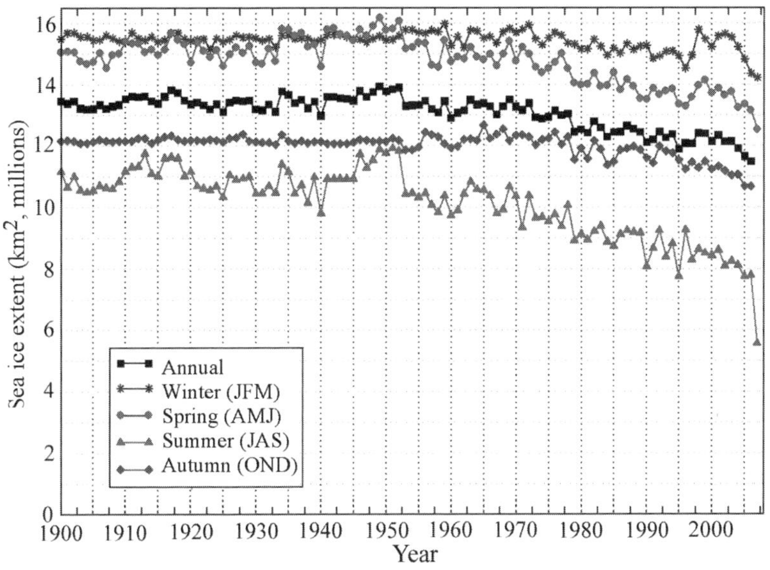

SOURCE: Cryosphere Today 2007. http://arctic.atmos.uiuc.edu/cryosphere/.
NOTE: km^2 = square kilometer.

that underlies Gore's image. This "new" set of old data sounds like the same Russian data set discussed by Ahlmann and more recently by Mahoney. When O. M. Johannessen and his colleagues, of Norway's Nansen Research Center, used the "hitherto under-recognized" Russian sea ice extent observations to create a long-term 20th-century record of sea ice, they produced an Arctic sea ice–extent history that looked quite different from the Gore version, and, in fact, exhibits a much higher correspondence to the Arctic temperature history (Figure 4.18).

The authors note that the Russian sea ice observations do not encompass the entire Arctic (they are missing about 23 percent of the total area, primarily along the coast of North America, including the eastern Chukchi Sea, the Beaufort Sea, and the Canadian Arctic straits and bays), and that the data are inadequate during World War II and the early postwar years. That probably explains the lack of correspondence between the Arctic sea ice extent and falling

Figure 4.18
ARCTIC SEA-ICE EXTENT RECONSTRUCTED FROM RUSSIAN DATA SET AND GORE'S LIKELY DATA SET, 1900–2000

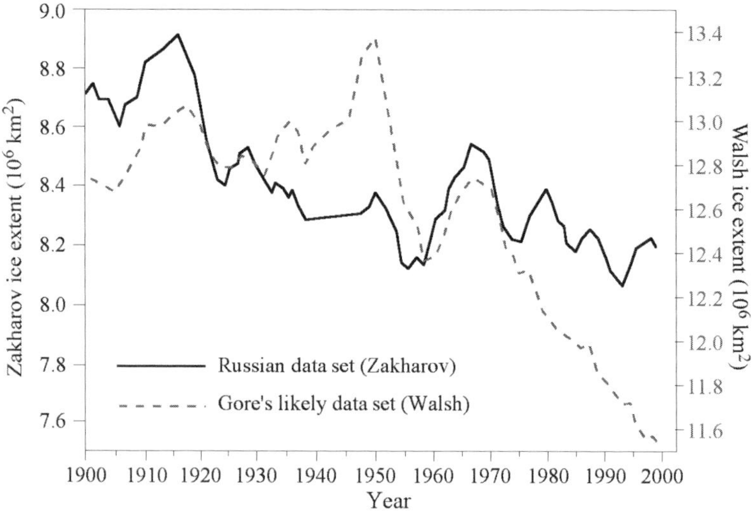

SOURCE: Adapted from Johannesssen et al. 2004.
NOTE: km^2 = square kilometer.

temperatures during the 1940s, as well as why the Russian reconstruction doesn't fall off as much in recent years (where a lot of sea ice loss took place off the northern coast of North America). But despite these difficulties, the Russian reconstruction shows far more interdecadal variation, including a large decline from 1900 to the 1940s, a recovery from the 1940s into the late 1960s (quite possibly underestimated due to insufficient data during the early part of that period), and then a subsequent decline to the present. The present decline has resulted in the absolute lowest sea ice extent area, but it has not progressed at the absolute fastest rate—which occurred early in the 20th century.

Mahoney explained in his seminar: "[T]he Russian Arctic ice pack did not fully recover during [the midcentury], suggesting that the early 20th century warming . . . may have preconditioned the Arctic for greater change in recent decades." In other words, human activity may be responsible for pushing Arctic sea ice to its lowest extent in the past 100 years or so, but we had quite a bit of help from Mother Nature. There has been a large degree of variation in Arctic sea ice extent over the course of the 20th century, much of which was fueled by non-human–induced climate variations.

We'll now progress a bit further to the South, into the permafrost and northern limits of tree growth. What's happening isn't exactly new.

Permafrost's Future

Anyone who Googles global warming will quickly discover that Arctic region permafrost is melting at an unprecedented rate, and somehow this will lead us to a runaway greenhouse effect that might warm the earth far more than any of us ever feared. The melting of permafrost is one of the main pillars of the global warming horror story.

There has been a flurry of recent research about the melting permafrost issue. If you do nothing more that search the Internet for "global warming + permafrost," literally hundreds of thousands of pages will pop up. You will read repeatedly that permafrost is a sink (storage mechanism) for carbon dioxide and another greenhouse gas, methane. When the soils of the high latitudes froze at the beginning of the last ice age, they trapped very large amounts of organic material (carbon-rich grasses, animal remains, and soil material) in frozen permafrost. As it thaws, then, the carbon trapped within the once-frozen soils will be released, mainly as methane, causing even more warming worldwide. We know that the warming is predicted to occur the most in the high-latitude land areas of the Northern Hemisphere—exactly where the permafrost is found today. With more warming in permafrost regions, more permafrost will melt, more methane will be released, and so forth.

The entire process is described by many as a time bomb that is going off before our very eyes. The bomb is not just causing the world to warm at a more rapid pace, but also the melting permafrost is also routinely connected to the destruction of forests, collapse of homes and other structures (e.g., pipelines), erosion of coastal areas and hillsides, disruption of animal habitat—and, well, you name it.

A very interesting article appeared in a 2007 issue of *Geophysical Research Letters* titled "Near-Surface Permafrost Degradation: How

Severe during the 21st Century?'' by G. Delisle from Hanover, Germany. Bet you haven't heard a word of it.

The final sentence of the abstract states: ''Based on paleoclimatic data and in consequence of this study, it is suggested that scenarios calling for massive release of methane in the near future from degrading permafrost are questionable.'' Delisle notes that warming is occurring in the Arctic regions, and that the warming will undoubtedly impact the permafrost of the high latitudes. Delisle notes, however, that many numerical models used to simulate the impact of warming on permafrost deal only with the upper 10 feet of the earth's surface. The previously used models do not take into account the cooling effect of deeper and colder zones that interact thermodynamically with the active layer near the surface. Delisle also exposes other assumptions of previous models that are ''in clear conflict with field evidence.''

Delisle presents ''a unidimensional long term permafrost temperature model of general application . . . which is capable to fully incorporate all relevant thermal processes within the active layer and the permafrost, and between the permafrost and the non frozen ground below.'' The model space is made up of 600 layers, with a minimum spacing of 10 cm (four inches) within the active layer and the uppermost ''permafrost zone.'' Rather than looking at only 10 feet beneath the surface as was done by previous models, the new model goes 100 yards into the surface, which provides a more realistic picture.

Using this more complete model, Delisle reports that continuous permafrost in Alaska and Siberia will persist over the next 100 years, even if a significant warming takes place. Further, we learn that

> Based on this result and on the presented analysis, it appears that all areas north of 60°N will maintain permafrost at least at depth. North of 70°N, surface temperature values today are in general below −11°C. These areas should maintain their active layer. It appears unlikely that almost all areas with near-surface permafrost today will lose their active layer within the next 100 years.

Delisle claims that the new model is far more consistent with field measurements and far more realistic in terms of including the interaction with the deeper and colder permafrost core.

Another common fear is that melting of permafrost will release trapped methane. Delisle notes this at the end of his article:

> A second, rarely touched upon question is associated with the apparently limited amount of organic carbon that had been released from permafrost terrain in previous periods of climatic warming such as e.g. the Medieval Warm Period or during the Holocene Climatic Optimum [the warmer millennia after the end of the recent ice age—see our next section]. There appear to be no significant [methane] excursions in ice core records of Antarctica or Greenland during these time periods which otherwise might serve as evidence for a massive release of methane.

A Long-Term History

What about the longer perspective? French climate researcher Jean-Claude Duplessy found that the Barents Sea—the portion of the Arctic Ocean bordering on Scandinavia and northwestern Russia—was 2°C (3.6°F) warmer 7,000 to 8,000 years ago than it was at the beginning of the Industrial Revolution. In the same era, Sigfus Johnsen of the University of Copenhagen found Greenland temperatures to be 0.5°C to 2°C (0.9°F to 3.6°F) warmer.

The most comprehensive analysis of Eurasian temperature histories back to the end of the last ice age was published in 2000 by Glen MacDonald, who chairs the Geography Department at UCLA.

MacDonald et al. collated records of trees preserved in the acidic environment that is now the Arctic tundra. The remains can be dated by radiocarbon analysis.

The boundary between the northern forest and the bare tundra is currently south of the Arctic Ocean, and is determined by summer maximum temperatures. MacDonald found: "Over most of Russia, forest advanced to or near the current arctic coastline between 9000 and 7000 yr B.P. [before present] and retreated to its present position by between 4000 and 3000 yr B.P."

In other words, the Eurasian arctic was considerably warmer than today for seven millennia!

How warm? "During the period of maximum forest extension, the mean July temperature along the northern coastline may have been 2.5 to 7°C [4.5°F to 12.6°F] warmer than modern [temperatures]."

One reason he gives for this warmth is "extreme Arctic penetration of warm North Atlantic Waters."

Imagine yourself on a satellite over the North Pole. What you will note is that there is only one "gate" through which such water can pass, and it is via the passage between Greenland and Europe. In other words, the east coast of Greenland was likely to have been warmer for *several millennia* and it did *not* shed its ice. James Hansen's speculation that a 1°C (1.8°F) rise in global temperature will therefore cause this is simply bad science fiction (as opposed to good sci fi, which is at least plausible). Further, the polar bear survived and the Inuit culture radiated.

Goodbye to three modern climate myths: Arctic temperatures are higher than any observed since before the last ice age; Greenland is about to shed all its ice; and global warming is driving the Inuit and the polar bear to extinction.

Postglacial warming was Arcticwide. Darrell Kaufman, of Northern Arizona University, noted that for 2,000 years—from 9,000 to 11,000 years ago, Alaskan temperatures averaged 3°F (1.7°C) warmer than now. Feng Sheng Hu, of the University of Illiniois, found that there have been three similarly warm periods in Alaska: 0 to 300, 850–1200, and 1800 to the present. Thompson Webb III, found timings similar to MacDonald: northwestern and northeastern North America were more than 4°F (2.2°C) warmer than the baseline from 7,000 to 9,000 and 3,000 to 5,000 years ago, respectively. And in a 2006 comprehensive review of regional temperature histories, the University of Buffalo's Jason Briner and colleagues wrote:

> . . . summer temperatures from Qipisarqo Lake on southern Greenland were 2°C to 4°C [3.6°F to 7.2°F] warmer in the early Holocene [post–ice age era beginning around 11,500 years ago] versus the late Holocene [more recent era]. . . . Greenland ice sheet borehole paleothermometry indicates a temperature change of ~3.5°C [6.3°F] between the middle and late Holocene [roughly 4,000 to 7,000 years ago].

. . . and ice did not fall off of Greenland, despite millennia of warmer temperatures.

Baked Alaska?

> "Global Warming is Killing Us, Too, Say Inuit."
>
> The Inuit people of Canada and Alaska are launching a human rights case against the Bush administration claiming they face extinction because of global warming.
>
> —*The Guardian*, December 11, 2003

Low-Balling Warming?

Obviously, MacDonald's paper is somewhat disturbing to global warming alarmists. Among other things, it surely argues that polar bears and Arctic people are pretty resilient, and that the Arctic can be warm for millennia and still recover its sea ice.

What's disturbing to us, though, is the way it was played in the IPCC process. As a reviewer of the 2007 "Fourth Assessment Report," one of us (Michaels) was privy to the various drafts that eventually became the final report.

Figure 4.19 (see insert) is from the "First Order Draft." The figure clearly shows "Siberia and East Russia" to be more than 2°C (3.6°F) warmer than the preindustrial era. That is a profound understatement, given that MacDonald noted summer temperatures 2.5°C to 7°C (4.5°F to 12.6°F) warmer than what he calls "modern." And the Alaskan ("Northwestern North America") temperatures should have been red, not orange, because Kaufman's work used the present rather than the colder "preindustrial" data as a baseline.

Between the first and second drafts, an amazing thing happened.

As shown in Figure 4.20 (see insert), the 7,000 years in which "Siberia and East Russia" were colored red (more than 2°C [3.6°F] warmer than preindustrial) simply *disappeared*, despite the reference to MacDonald et al. in the accompanying figure caption.

Figure 4.21 (see insert) is from the final, published version, which appeared in May 2007. "Siberia and East Russia" from Figure 4.19 (absent in Figure 4.20) have now been replaced by "North Eurasia," with the warm period ending 8,000 years ago, instead of 10,000. Contrast that with Macdonald's text:

> Radiocarbon-dated macrofossils are used to document Holocene [post-ice age] treeline history across northern Russia (including Siberia). Boreal forest development in this region commenced by 10,000 yr B.P. Over most of Russia, forest advanced to or near the current arctic coastline between 9000 and 7000 yr B.P. and retreated

(continued on next page)

(continued)

> to its present position by between 4000 and 3000 yr B.P. Forest establishment and retreat was roughly synchronous across most of northern Russia.

All of these shenanigans certainly provoke a number of questions. Why did the entire warming disappear from the second draft, why did it appear in a form that was obviously not what the author intended in the final draft, and why didn't the IPCC mention how much warmer it actually was, rather than dialing it down to simply "above 2°C [3.6°F]"?

Really?

The Alaska Climate Research Center, at the University of Alaska–Fairbanks, maintains the statewide temperature database along with historical analyses. According to the center's website (http://climate.gi.alaska.edu), "[T]he period 1949–1975 was substantially colder than the period 1977–2003; however, since 1977 *no additional warming has occurred in Alaska* [emphasis added] with the exception of Barrow and a few other locations."

In 1976, a stepwise shift appears in the temperature data, which corresponds to a change in the distribution of Pacific Ocean temperatures. Commenting on this shift, in 2005 Brian Hartmann and Gerd Wendler wrote in the *Journal of Climate*:

> The regime shift [was] also examined for its effect on long-term temperature trends throughout the state. The trends that have shown climatic warming are strongly biased by the sudden shift from the cooler regime to a warmer regime in 1976. When analyzing the total time period from 1951 to 2001, warming is observed, *however the 25-year period trend analyses before 1976 (1951–75) and thereafter (1977–2001) both display cooling* [emphasis added].

This behavior is certainly contrary to what is produced by climate models forced with increasing carbon dioxide. The warmings those models calculate are generally smooth in character. None of them predicts a sudden one-year shift in Pacific temperatures, followed by a quarter-century of stable temperatures in nearby land regions.

Of course, it *could* be the fact that greenhouse warmings do in fact manifest themselves largely as a series of fits and starts, but if that is true, then the models are currently incapable of accurately simulating the response of regional temperature to changes in global carbon dioxide.

Vanishing Alaska?

The *Times* and other big papers have recently published a tremendous number of articles about changes in the Arctic, noting that there have been massive changes, including erosion along the sparsely settled north coast of Alaska, presumably caused by global warming; Figure 4.22 (see insert) shows a photograph that accompanied one such article. In fact, there's little "news" here that is fit to print. It's a very old story. Alaskans were warned of this almost a half-century ago.

An intense storm struck the northwestern tip of Alaska during the fall of 1963, causing over $3 million in damage in the much more valuable dollar of the time. A 10-foot storm surge (which would be respectable for a tropical hurricane) gravely damaged a U.S. government research camp that was located at Barrow as winds gusted to hurricane force. The storm hit during an unusual ice-free period in early October—the primary reason why the seas grew to such damaging heights. During most months of the year, near-shore sea ice coverage is sufficient to dampen (or prevent entirely) the buildup of significant wave heights.

James Hume (at Smith College) and Marshall Schalk (then of Tufts University) described the damage from the 1963 storm in an article written for the journal *Arctic* in 1967. On the basis of historical weather records and the recollection of Inuit elders, they estimated that it was about a "200-year" storm at this location.

This storm, and others like it (it's easy to have many "200-year" storms in a few years, because they can strike in different places), should have served as ample warning against settling on the unstable coastline of much of Alaska.

The wind and waves from the great 1963 storm took a huge toll on the Barrow shoreline. Hume and Schalk estimated the erosion damage from the single 1963 storm to be equivalent to about 20 years of "normal" erosive processes. And the "normal" erosive processes themselves were known to be substantial along much of

Alaska's coast, which is made up of loose sediments held together by ice. Erosion rates have been measured to range from a few feet to a few *tens of feet per year* along much of Alaska's western and northern shorelines.

A. D. Hartwell, then at the U.S. Army's Cold Regions Laboratory in Hanover, New Hampshire, described the processes acting on the northern Alaska coast in a 1973 paper in *Arctic:*

> Most of this coastline is marked by an abrupt break in slope between the relatively horizontal terrain of the mainland and the gently-sloping sea floor. In bedrock areas this break is generally a steep sea cliff with loose talus material at its base. In areas of perennially frozen sediment which are exposed to direct wave attack along the coast, the relief is often sheer and is formed by slumping of large blocks of frozen sedimentary material. This is a result of both thermal and mechanical erosion along the base of the sea cliff and inland along the banks of estuaries and rivers where undercutting of the frozen sediments forms a "thermoerosional niche." Such niches which are unique to this environment can form rapidly and may extend several metres under the bank, making the overhanging bank unstable and susceptible to collapse especially where ice wedges are intersected.

Acting on top of those erosive processes are strong late summer and early fall storms, such as the one in October 1963, After comparing the high rates of event-based erosion (such as the 1963 storm) with the ongoing long-term erosion rates, the Hume and Schalk paper ended with an eerie warning, in 1967, about the future:

> A practical consideration also arises from this study. If, as has been suggested, the climate is becoming warmer as a result of the addition of carbon dioxide to the atmosphere (Plass 1956; Callender 1958; Kaplan 1960; Mitchell 1965), the likelihood of an open ocean and strong winds coinciding to produce such a storm in the future is constantly increasing. Another such storm can be expected, and care should be exercised in the selection of building sites and construction methods. The best sites would be at least 30 feet above sea level and either inland or along a coast which is not eroding.

Much of this advice went unheeded. Nowadays, we hear story after story describing the plight of the native Alaskans as their villages, which were constructed on the unstable bluffs along the Alaskan coast, are undermined by the retreating shoreline.

As native Alaskans began a transition from their traditional nomadic lifestyles to more permanent villages, replacing snow houses with tin and plywood buildings, dogsleds with snowmobiles, and seal-oil lamps with electric lights, they located many of those settlements very near the (already receding) shoreline to provide ready access to the oceans, a primary source of the coastal Inuit's sustenance.

In earlier times, when the Inuit were more nomadic, they simply would have broken camp and moved to a more suitable location. In fact, the historical scientific literature contains references to abandoned Inuit camps located on the precipices of an eroding coast. Gerald MacCarthy, then at University of North Carolina, in an article published in *Arctic* in 1953 titled, "Recent Changes in the Shoreline near Point Barrow, Alaska," wrote:

> At "Nuwuk" [also called "Newtok," the same location photographed by the *New York Times* in 2007; see Figure 4.22 in insert] the evidence of rapid retreat is especially striking. The abandoned native village of the same name, which formerly occupied most of the area immediately surrounding the station site, is being rapidly eaten away by the retreat of the bluff and in October 1949 the remains of four old pit dwellings, then partially collapsed and filled with solid ice, were exposed in cross section in the face of the bluff. In 1951 these four dwellings had been completely eroded away and several more exposed.

What's new here? Not much. Hume et al., in a 1972 paper, include a 1969 photograph with the caption: "Aerial view of the bluffs near the village recently settled. One building collapsed and one has been moved from the bluffs as a result of the 1968 storm. The beach formerly was 30 m. in width at this point. Photo taken in August 1969." The authors go on to add, "The village will probably have to be moved sometime in the future; when depends chiefly on the weather. . . ." (We do not reproduce the picture here because it is of very poor quality.)

Clearly, erosion has been gnawing away at the Alaska coast for many, many decades and this fact has been known for a long time. Wind and waves acting on soil held together by ice acts through a destructive feedback to expose more frozen soil to the above-freezing temperatures of summer and the warm rays of sunshine, softening it for the next round of waves and wind. And so the process continues. A decline in near-shore ice cover helps to exacerbate the process. Ignoring these well-known environmental conditions has led to the unfortunate situation today where Inuit villages are facing an imminent pressure to relocate. This situation has less to do with anthropogenic climate change than it does with poor planning in the light of well-established environmental threats, with or without global warming.

Antarctica

Antarctica's ice sheets and glaciers are the largest mass of ice on the planet, comprising some 25.71×10^6 cubic kilometers (6.18×10^6 cubic miles), or 89.5 percent of total global land ice. Global warming theory predicts, in general, that warming is enhanced in cold, dry regions, but Antarctica is an exception. There's plenty of evidence that, as a whole, it hasn't warmed a bit in the last four decades, or even may have cooled, as shown by Peter Doran in *Nature* in 2002.

Doran's work was extensively noted in *Meltdown*. He found a net cooling since 1966, but a strong warming around the small Antarctic Peninsula, the strip of land comprising less than 2 percent of the continent that juts out toward South America.

There are more recent analyses. In 2007, William Chapman and John Walsh of the University of Illinois extensively reviewed and updated the climate literature on Antarctic warming, concluding "These studies are essentially unanimous in their finding that the Antarctic Peninsula has warmed since the 1950s, when many of the surface stations were established." But that's hardly the true picture of what is happening over the continent as a whole. They wrote, "Recent summaries of station data show that, aside from the Antarctic Peninsula and the McMurdo area, one is hard-pressed to argue that warming has occurred, even at the Antarctic coastal stations away from the peninsula and McMurdo." Furthermore, they state, "Recent attempts to broaden the spatial coverage of temperature estimates have shown a similar lack of evidence of spatially widespread warming."

Chapman and Walsh collected temperature in and around Antarctica from 460 locations including 19 manned surface observation stations located on the continent, 73 automated weather stations, and a 2°-latitude-by-2°-longitude gridded sea-surface temperature time series. They made every attempt to have complete records from 1958 to 2002.

When averaged over the entire region from 60°S to 90°S (an area much larger than Antarctica proper), Chapman and Walsh found:

> The 45-yr linear temperature change is largest in winter (+0.776°C; 1.40°F) and spring (+0.405°C; 0.73°F), and smallest in summer (+0.193°C; 0.35°F) and autumn (+0.179°C; 0.33°F). These temperature changes correspond to linear trends of +0.172°C/decade; 0.31°F (winter), +0.090°C/decade; 0.16°F (spring), +0.045°C/decade; 0.08°F (summer), and +0.040°C/decade; 0.07°F (autumn).

But these are for the whole region, rather than the continent. Referring specifically to Antarctica, they found that "the 45 yr (1958–2002) linear temperature change of annual mean temperatures is +0.371°C (0.66°F) with a corresponding trend of +0.082°C per decade (0.14°F)." Furthermore, the authors found: "Statistically significant warming is confined to the Antarctic Peninsula and a small region along the eastern coast of the continent. Temperature trends over the remainder of the Antarctic continent *do not exceed significance thresholds* [emphasis added]." So, at this point, we have learned that Antarctica is warming in its bitterly cold winter season and most of the warming is confined to a small area surrounding the Antarctic Peninsula (Figure 4.23; see insert).

Here is the interesting twist to the story. The results are actually quite consistent with Doran's finding of a cooling (noted in *Meltdown*), given that the Doran study began in 1966. Graphs of seasonal and annual temperature trends show that the coldest years tend to occur at or near the beginning of the record, in the 1950s. Chapman and Walsh find: "Trends computed using these analyses show considerable sensitivity to start and end dates with starting dates before 1965 producing overall warming and starting dates from 1966 to 1982 producing net cooling rates over the region." In this case, "region" refers to the entire area south of 60° latitude.

Because a cooling or a constant-temperature Antarctica seems so counter to greenhouse theory, it's necessary to come up with some logical construct to explain its behavior.

There are two competing hypotheses. The first involves another set of industrial emissions, the chlorofluorocarbons (CFCs). CFCs were originally used as inert propellants in aerosol spray cans, and as refrigerants. They haven't been seen in a spray can since the late 1970s (contrary to what many people think), and they were phased out as refrigerants via the Montreal Protocol, a UN agreement that went into force in 1989. (The protocol actually allows developing countries to produce CFCs through 2010, but there has already been a slight reduction in their overall atmospheric concentrations.)

CFCs break down into their constituent elements, including chlorine and fluorine, which have long been known to assist in the destruction of ozone under conditions observed in the stratosphere. Ozone absorbs ultraviolet radiation from the sun (the same type of radiation that gives us sunburn), and therefore it keeps the stratosphere warmer than it would be if it were absent.

Over most of the planet, the stratosphere begins some 8 to 10 miles above the surface. But, in frigid Antarctica, it lies much closer, as the coldness of the atmosphere compresses its layers. The stratosphere gets down to about 25,000 feet above sea level at the South Pole, which itself is at 9,300 feet. The distance between the stratosphere and the Pole is therefore roughly 16,000 feet (on the U.S. East Coast, the distance is about 35,000 feet), so the unusually cold stratospheric air, chilled further by ozone depletion, occasionally mixes down and is found over the entire continent. Hence, a cooling trend.

A second (and not necessarily competing) explanation is that the lack of warming is *caused* by warming of the surrounding ocean.

Huh? Most people might lump anyone who says this with our neighbors who insist that if you put hot water in the ice-cube tray, then it freezes faster than cold water. That's just not so and violates physical thermodynamics.

But special factors come into play owing to Antarctic geography. The surrounding ocean has warmed a few tenths of a degree (Celsius). That might not seem like much, but even a teensy warming of water increases the amount of water vapor that is given up to the atmosphere, and the vastness of the Southern Ocean will give up a lot of molecules even if it warms only a bit. Northerly winds, which are common, will push this moisture up and over the continent.

Antarctica (or any continent) is a lot rougher than the surrounding smooth ocean, so any air stream that impinges upon it has to slow down. That phenomenon, known as "convergence," leaves the air nowhere to go but up (it can't go down into the continent!) and this results in the formation of clouds, as the air ascends, cools, and reaches the level at which clouds condense out.

Clouds of this derivation are pretty shallow—usually less than 10,000 feet in height—and bright white when viewed from above. As a result, they reflect away much of the sun's energy—much more than they absorb from the earth because of their extensive amount of water vapor. Consequently, they could produce a net cooling of the continent.

Further, they should produce more snow. In fact, computer models for 21st-century climate have Antarctica gaining ice—that is, contributing to a relative *lowering* of sea-level rise, because of an increase in snowpack.

The two possible mechanisms for Antarctic cooling—stratospheric ozone loss and increased reflective cloudiness caused by a warming ocean—aren't going to go away soon.

Long-term studies of Antarctic ice differ. A paper published in *Science* in 2005 by the University of Missouri's Curt Davis and colleagues indicated a net increase in ice mass over the continent (Figure 4.24). Though there were declines in ice over the Antarctic Peninsula and the adjacent West Antarctic ice sheet, there were gains in the much larger East Antarctic ice sheet. A highly cited much shorter-term study, written by Isabella Velicogna and John Wahr and published in *Science* in 2006, found a decline in the West and no net change in the East (Figure 4.25).

Note that in the Davis study that the East reaches its maximum elevation in mid-2002. This is the same time as the starting point for the Velicogna and Wahr study, where the first data point, in mid-2002, is also the highest. The Velicogna and Wahr study is necessarily short because the satellite measuring system, called GRACE ("Gravity Recovery and Climate Experiment") satellite only became operational in early 2002. If one puts the two together, it still looks like a net increase for the East.

Andrew Shepard and Duncan Wingham, two British scientists, summarized six recent studies of Antarctic ice and concluded that it is most likely that the East Antarctic ice sheet is gaining about 6 cubic miles of ice per year, and the West Antarctic losing about 12, for a net loss of 6. They argue that the data are so poor that it is

Antarctica: More Fact-Checking, Please

Sure enough, one can use a computer to combine the effects of ozone depletion and the patterns of warming in the far Southern Hemisphere to forecast the future, which is what NASA's Drew Shindell and Gavin Schmidt did in *Geophysical Research Letters* in 2004.

Without being technical to the point of boredom, it has been the "normal" condition for temperatures around Antarctica and those on the continent to change in different directions. Scientists even give this a name, the "Southern Annular Mode" (SAM), "Annular" referring to its ringlike structure, owing to the geometry of the ocean surrounding Antarctica.

Their computer model predicts that SAM is going to pretty much disappear because of global warming. NASA's press office then produced a lurid press release (!) about Shindell and Schmidt's modeling results, promising certain disaster for the region because of "ice sheets melting and sliding into the ocean" leading to "greatly increasing sea levels." It might be worth noting that James Hansen is Gavin Schmidt's boss.

Only one problem, which is more than vaguely analogous to the difficulties with Hansen's hypothesis that Greenland is about to fall apart: This study, too, does not comport with history.

Shindell and Schmidt claim that SAM is in its current position because of stratospheric ozone depletion over Antarctica, which, they say will become much less significant in coming decades as the putative cause—CFCs—are phased out.

But J. M. Jones and M. Widmann showed in *Nature* in 2004 that the SAM looked a lot like it does today some 40 years ago—long before ozone loss. Further, they found it resembled what NASA forecasts for the future, even though the planet was *cooler* during the first half of the 20th century!

difficult to assume that there have been any major changes in the last decade. Nor are the mechanisms clear. It's clearly too cold to be caused by melting. For example, the Amundsen Sea, in which the West Antarctic ice sheet terminates, shows no evidence for warming.

Figure 4.24
ELEVATION CHANGE OF THE EAST ANTARCTICA ICE SHEET, 1992–2003

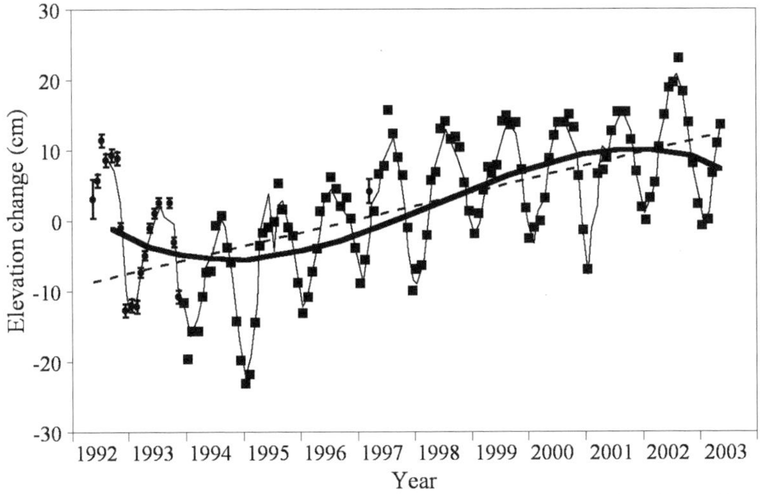

SOURCE: Davis et al. 2005.
NOTE: cm = centimeter.

As noted earlier, Shepard and Wingham also summarized recent findings for Greenland's ice and estimated a net annual loss of 25 cubic miles. When combined with the data from Antarctica, the loss figures contribute to an annual sea-level rise of 0.01 inches per year, an amount far too small to measure. Although they express concern that current models for ice dynamics are very crude, oddly enough, they make no reference to the long period in the early 20th century when Greenland was warmer and that condition obviously triggered no catastrophic change in the behavior of the 685,000 cubic miles of ice that sit atop the island continent.

Antarctic Paradox: Does Less Equal More?

> "Escalating Ice Loss found in Antarctica: Sheets Melting an Area Once Thought to Be Unaffected by Global Warming"
>
> —*Washington Post*, January 14, 2008

In a multipage article beginning above the fold on the front page, *Post* writer Michael Kaufman describes the just-published research

Figure 4.25

ICE-MASS ESTIMATE BY THE GRACE SATELLITE FOR WEST ANTARCTIC AND EAST ANTARCTIC ICE SHEETS, 2002 TO MID-2005

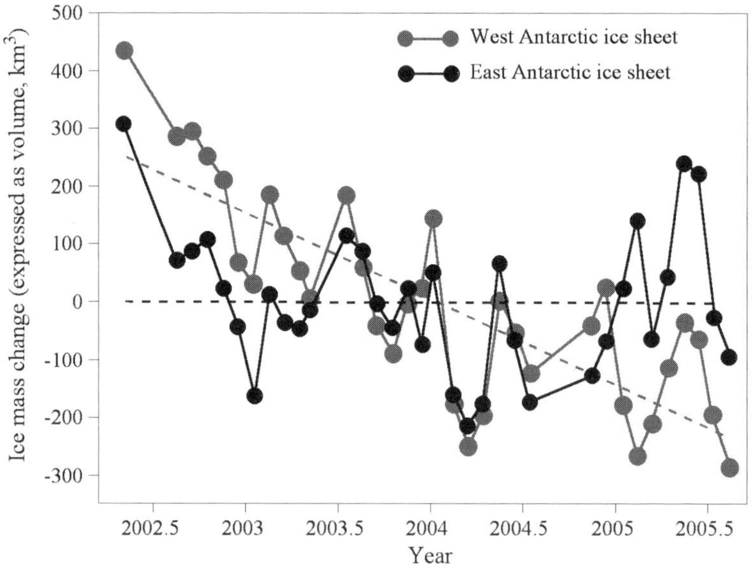

SOURCE: Velicogna and Wahr 2006.
NOTE: km^3 = cubic kilometer.

results by Eric Rignot and colleagues: "Climatic changes appear to be destabilizing vast ice sheets of western Antarctica that had previously seemed relatively protected from global warming . . . raising the prospect of faster sea-level rise than current estimates."

Rignot et al.'s study came out after the summary article of Shepherd and Wingham. But in reality, it's merely consistent with many of the studies noted there.

Rignot et al.'s study is at variance with all recent simulations of 21st-century climate in the Antarctic, which predict a gain in ice because of increasing snowfall. Because of this conflict, and because of clear indications of a net gain in the entire Southern Hemisphere ice extent (see below) it's currently impossible to say what is really happening.

Rignot et al. use satellite observations to determine ice stream velocity and ice thickness, which are combined to calculate out how much ice is flowing off Antarctica and into the oceans. This represents the total ice loss from Antarctica, as it is generally much too cold for melting to be a significant source of ice loss. To determine ice gains—through snowfall accumulation—the researchers used a weather model to *simulate* snowfall. The net change in ice is the difference in the input minus the outflow. Their study covered the period 1992–2006.

Why use a model of ice input (snowfall) and actual observations of outflow? Well, actual snowfall measurements are few and far between over the vast continent of Antarctica, and it's very hard to measure, because much "new" snow is admixed with old stuff blown off of the surface. Rignot et al. do not use the modeled *annual* values of snowfall, but instead use the average modeled snowfall across the years 1980 through 2004. Rignot calculates actual outflow for various years from the satellite observations, but uses a fixed amount of input (i.e., snowfall) that represents average conditions rather than the year-to-year variations.

In reality, there is a tremendous amount of interannual and interdecadal variation in snowfall across Antarctica. In 2006, Andrew Monaghan, of Ohio State University's Byrd Polar Research Center, and colleagues examined the snowfall history over Antarctica from 1955 through 2004 (again using snowfall amounts produced by weather models and verified by the available observations—mostly ice cores rather than direct snowfall observations). They concluded that there has not been any appreciable change in snowfall over Antarctica over the full period of record of their study, but that there is a fairly large year-to-year and decade-to-decade variation.

Again, it is worth noting that all modern climate models predict that Antarctica will gain mass as the climate warms because the continent will see an increase in snowfall—enough to offset glacial ice losses along the periphery. But, as Monaghan et al. recently reported, this snowfall increase has not been detected. Does this represent a failure of all climate models? If this is indeed true, what does this leave us for future projection?

Kaufman closed his *Post* article by noting that the head of the Intergovernmental Panel on Climate Change, Rajenda Pachauri, was heading down to Antarctica that week "to get a firsthand view of the situation."

If the weather was clear, before he even got there he was in for a shock. Despite an overall slight warming of the Southern Ocean,

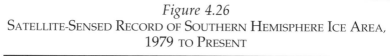

Figure 4.26
SATELLITE-SENSED RECORD OF SOUTHERN HEMISPHERE ICE AREA, 1979 TO PRESENT

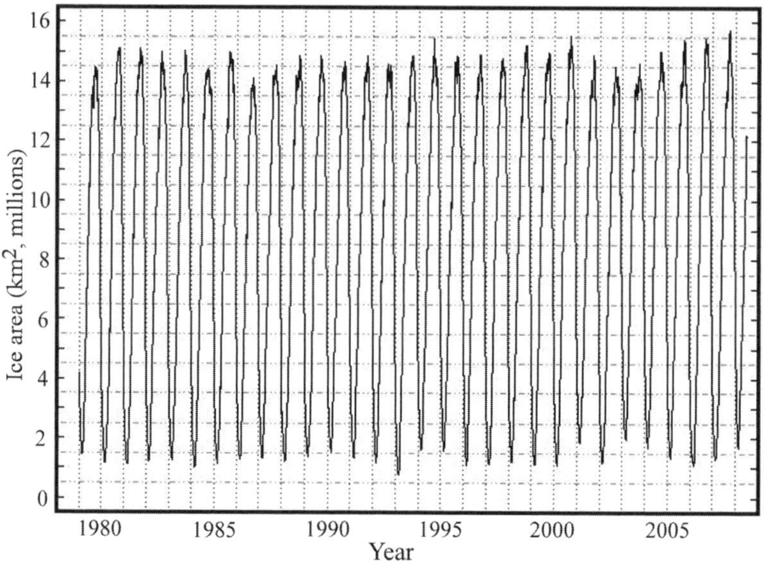

SOURCE: Cryosphere Today 2007. http://arctic.atmos.uiuc.edu/cryosphere/.
NOTE: km^2 = square kilometer.

the amount of ice surrounding Antarctica reached all-time record levels in 2007. Satellites first began to monitor this almost 30 years ago. Figure 4.26 shows that the ice reached its greatest extent in southern winter 2007 (northern summer), and that departure from average for a given month, in January 2008 (when the *Post* article came out), was the greatest ever measured (Figure 4.27).

Midway through the very long January 14 article is a statement that "these new findings come as the Arctic is losing ice at a record rate." Wouldn't that have been the appropriate place to note that the Southern Hemisphere was, at the same time, setting records for overall ice extent? (As of this writing, in September 2008, the Southern Hemisphere ice anomaly is back to its normal range.)

Kilimanjaro Redux

> The "snows" of Africa's Mount Kilimanjaro inspired the title of an iconic American short story, but now its dwindling

Figure 4.27
SOUTHERN HEMISPHERE SEA ICE ANOMALY, 1979–2008

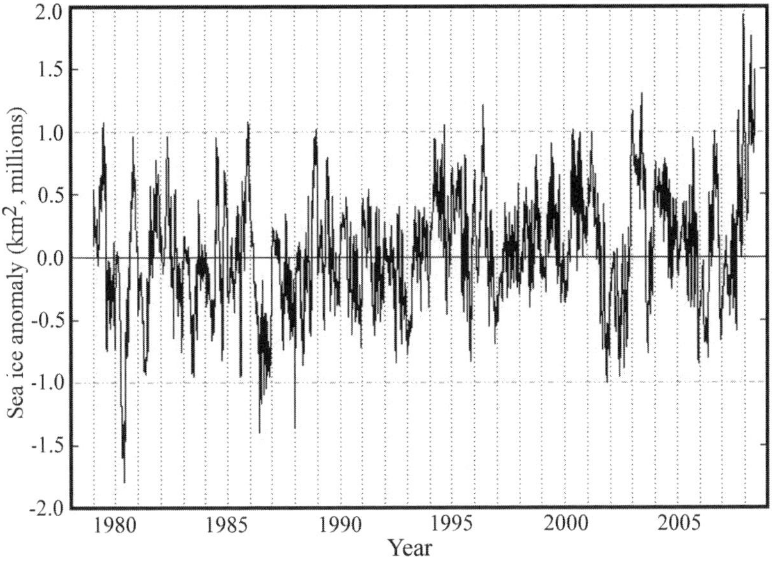

SOURCE: Cryosphere Today 2007. http://arctic.atmos.uiuc.edu/cryosphere/.
NOTE: km^2 = square kilometer.

> icecap is being cited as proof for human-induced global warming.
>
> However, two researchers writing in the July–August edition of *American Scientist* magazine say global warming has nothing to do with the decline of Kilimanjaro's ice, and using the mountain in northern Tanzania as a "poster child" for climate change is simply inaccurate.
>
> —University of Washington Press Release, June 11, 2007

Meltdown led off its parade of climate horrors with "The Snowjob of Kilimanjaro," demonstrating that, if it were 1976 (after three decades of global cooling), one could have written "Kilimanjaro's glaciers will completely disappear by 2015 if this cooling trend continues."

What's new since *Meltdown*?

Al Gore featured Kilimanjaro in his book and movie, giving it a four-page spread in the print version, and featuring Ohio State

University's Lonnie Thompson, a scientist and chief publicist for the Kilimanjaro-Will-Disappear-Soon story, standing next to the "pitiful last remnants of one of [Kilimanjaro's] glaciers."

Really?

One nice thing about glaciers is they leave records of where they were, in piles of debris, called moraines, and these can be dated. It is well known that Kilimanjaro's glaciers were far advanced beyond where they are today when the earth was warmer, for several millennia after the end of the last ice age. So the glacier can obviously take (and prosper under) warmer conditions.

The old adage "it's not the heat, it's the humidity!" clearly applies to Kilimanjaro.

In 2004, Georg Kaser and four colleagues wrote in the *International Journal of Climatology*, that "a drastic drop in atmospheric moisture at the end of the 19th century and ensuing drier climatic conditions are likely forcing glacial retreat on Kilimanjaro." Overall, Kaser et al. wrote that ". . . the climatic evolution of East Africa . . . is characterized by a drastic dislocation around 1880, when lake levels dropped notably and glaciers started to recede from their maximum extent."

That is concurrent with the end of the "Little Ice Age," a cold period noted in many locations around the planet that lasted about 400 years. It also is near the point in time when surface temperatures around the planet began to climb, but long before there could have been much influence from increased carbon dioxide.

Kaser et al. say that the glaciers will survive, despite Gore's protestations. "If the present precipitation regime persists," they conclude, "then these glaciers will most probably survive in positions and extents not much different than today. This is supported by the [fact] that slope glaciers retreated more from 1912 to 1953 than since then."

In a 2006 edition of *Geophysical Research Letters*, N. J. Cullen et al., of the Tropical Glaciology Group at University of Innsbruck, should have finally stilled the Kilimanjaro hue-and-cry (but alas, did not). Cullen et al. point out that the glaciers on the mountain are above (higher than) the mean freezing level, meaning that it is "difficult to suggest that air temperature changes alone are responsible for glacier recession on Kilimanjaro."

To re-evaluate possible causes of glacier retreat on Kilimanjaro, Cullen et al. employed recent high spatial-resolution satellite images of the mountain to construct a new detailed map of the ice bodies. They compared the new data to long-term variations in ice extent to assess its retreat in the context of 20th-century changes in air temperature, atmospheric moisture, and precipitation in East Africa.

The group of researchers found that glacial retreat during the 20th century was profound, as their work shows that only 21 percent of the 1912 ice cover on Kilimanjaro existed in 2003 when the satellite images were taken. However, the highest recession rates occurred in the first part of the 20th century, while the recession rate over the last 15-year interval (1989–2003) was smaller than in any of the other defined intervals in the study period of 1912–2003. Obviously, this is counterfactual to the notion that the recession is largely caused by recent (anthropogenic) warming. Given this curious finding, Cullen et al. set out to interpret the findings in the context of 20th-century climate change.

The ice bodies of Kilimanjaro are stratified into two types of glaciers—plateau (elevation ≥5700 m [18,700 feet]) and slope (<5700 meters)—to help differentiate by physical features such as shape, slope, thickness, and bed shape. Characterized by name, plateau glaciers are tabular-shaped ice bodies that rest stably on flat surfaces near the top of the mountain. In contrast, slope glaciers are found on steeper surfaces and move downward. The retreat of all plateau glaciers was found to be continuous and linear since 1912, whereas slope ice bodies experienced a rather rapid recession between 1912 and 1953, followed by a decreasing rate of retreat. The constant retreat of the plateau glaciers does not indicate that climate fluctuations during the 20th century affected their demise. Instead, solar radiation incident on vertical walls of the glaciers produces irreversible melting despite air temperatures that remain below freezing. Cullen et al. conclude that the demise of the plateau glaciers of Kilimanjaro is unavoidable given their geometry, and that any recent change in climate has had no significant impact. Thus, the decline of the plateau glaciers of Kilimanjaro are excused from the global warming debate.

Unlike with plateau glaciers, Cullen et al. believe that slope glaciers have a much shorter adjustment time to changes in climate—on the order of a few years. The research group believes that the rapid recession in the early part of the 20th century (1912–53) indicates that the glaciers were wildly out of equilibrium in responding to the prior shift in climate during the late 19th century. Cullen et al. contend that the magnitude of the shift is so large that the slope glaciers are still out of equilibrium. They note there is no evidence for any atmospheric warming in the 20th century in the vicinity of the glaciers. Joining plateau glaciers, the slope glaciers of Kilimanjaro are excused from the global warming debate.

Cullen et al. conclude that "Rather than changes in 20th century climate being responsible for their demise, glaciers on Kilimanjaro appear to be remnants of a past [19th-century] climate that was once able to sustain them."

Snowpack in the Andes

Halfway around the world from Kilimanjaro run the spectacular Andes Mountains, the largest and most impressive range in the Americas.

Meltdown featured a 2001 *Washington Post* story about the decline and fall of Peru's glaciers, which we found odd, because we couldn't find any net temperature change for the last three decades. (Hmm . . . that sounds a lot like Kilimanjaro).

Outside Antarctica, snow cover in the Southern Hemisphere has not received much attention in the climate change debate. In fact, within the snow, ice, and frozen ground chapter of the 2007 IPCC report, approximately 800 words along with three figures and one table are dedicated to snow cover variability in the Northern Hemisphere, compared with less than 400 words and no accompanying graphics for variability in the Southern Hemisphere.

Very late in 2006, the *Journal of Climate* published a paper by Mariano Masiokas (Instituto Argentino de Nivología, Glaciología y Ciencias Ambientales and University of Western Ontario geography department) and colleagues titled, "Snowpack Variations in the Central Andes of Argentina and Chile, 1951–2005: Large-Scale Atmospheric Influences and Implications for Water Resources in the Region." The research team used snow data from each side of the central Andes in Chile and Argentina to develop the "first regional snowpack series." They examined the six longest and most complete snow records for the 55-year period in the region, covering an area stretching from 30°S to 37°S latitude. Their variable for study is annual maximum snow water equivalent (MSWE).

Figure 4.28
REGIONAL SNOWPACK (MSWE) FROM THE CENTRAL ANDES, 1950–2005

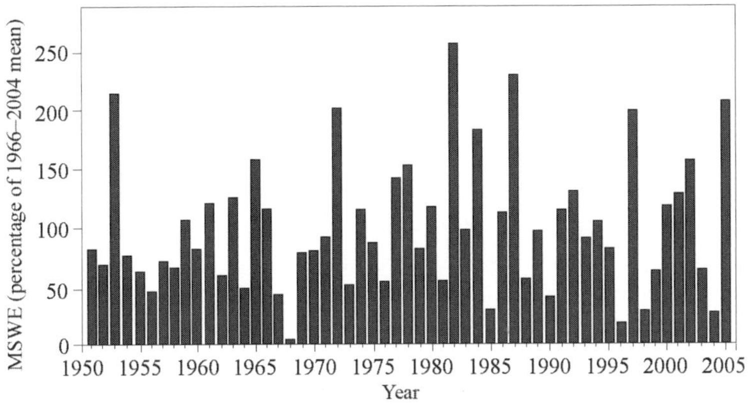

SOURCE: Adapted from Masiokas et al. 2006.

The snowpack of the central Andes serves as much more than a monitor of climate change. The authors explain that "over 10 million people in Central Chile and central-western Argentina depend on the freshwater originating from the winter snowpack of the central Andes." Alarming is their charge that "coupled atmosphere–ocean general circulation models especially targeted to investigate high-elevation sites" have indicated that "for the next 80 years the central Andes will probably experience significant temperature increases." To make matters worse, Masiokas et al. note, "independent general circulation model simulations also predict a significant decrease in precipitation over the region for the next five decades." The combination of higher air temperature and less precipitation in the central Andes over the rest of this century is not the recipe for a problem-free regional water supply. Climate models seem to be sending a strong message to the more than 10 million people in Chile and Argentina.

Masiokas et al. found no such trend in MSWE, stating that the regional record "shows a nonsignificant positive linear trend (+3.95 percent per decade) over the 1951–2005 interval," or an absolute increase of greater than 21 percent over the period (Figure 4.28). The group matched the MSWE record with mean monthly streamflow data for the primary rivers in the region. They found that river discharges on both sides of the central Andes "are strongly correlated with the snowpack record and show remarkably similar interannual variability and trends." In other words, the water supply is hardly decreasing.

In conclusion, there are a number of very interesting papers in the refereed literature revealing that the ice, snow, and sea-level rise story is a very complicated one. The latest compilations indicate an extremely small contribution of Greenland and Antarctica to sea-level rise, with little evidence for any marked change in the past decade. The causes of any putative loss in Antarctica are simply unknown, but one thing is for certain—it hasn't warmed up down there. Horror stories about an imminent collapse of Greenland's ice simply aren't borne out by the fact that it was warmer there for decades in the early 20th century, and for *millennia* after the end of the last ice age. And, finally, breathless stories about the end of the glaciers of Kilimanjaro and a decline in Andean water supply turn out to be snow jobs.

5. Extreme Climate: Floods, Fires, and Droughts

The Rest of the Storms

Although hurricanes are of special merit owing to their mediagenicity and their destructive potential, there are several other types of cyclones (low pressure areas) on this planet. The generic name for hurricane is "tropical cyclone," but their nontropical cousins (called extratropical cyclones) are orders of magnitude more common. In fact, hurricanes are quite rare.

Prove this for yourself by watching some cable weather channel. You'll see the maps dominated by nameless low pressure systems in the mid- and high latitudes. The occasional named tropical cyclone obviously commands a lot of attention ("Category 1 Hurricane Blowhard is about to demolish Cocoa Beach! We'll update you on this every two minutes!"—i.e., between commercials.) There are about a thousand more extratropical storms than hurricanes every year.

Some can be pretty destructive. In the last chapter, you read how a 1963 cyclone in Alaska caused dramatic erosion along the North Coast, and prompted a warning from scientists to build permanent structures far inland.

In October 1962, the Columbus Day storm brought wind gusts to 100 miles an hour in the Pacific Northwest, resulting in massive blowdowns of the extensive regional forest. In April 1974, a huge cyclone in the Midwest spun the atmosphere so hard that a still-record 148 twisters touched down from that single storm. Six were Category 5 tornadoes, an unheard-of number for one day.

Those storms occurred before the warming that started in the mid-1970s.

On March 13, 1993, a huge low-pressure system spun up in the Gulf of Mexico and exploded up the East Coast, earning the title "Storm of the Century," because it set many records for lowest barometric pressure ever measured at inland locations. (Note that hurricanes routinely have lower pressures, but that their barometric pressures rise as soon as they come ashore).

Kevin Trenberth, of the U.S. National Center for Atmospheric Research, went on *Meet the Press* the next day and blamed global warming.

Every strong European cyclone in the last decade has prompted a similar outcry. If there's a big storm, a reporter *will* find an ''expert'' who will conflate the wind with global warming. Just Google ''news'' after the next one to prove this to yourself.

All of this seems a bit illogical. Although hurricanes are, in part, driven by the heat of the ocean, there's a pretty strong debate, noted in chapter 3, about their relation to global warming. But the mechanism that creates and feeds extratropical cyclones is a lot different. They're driven by the jet stream, a circumpolar vortex of high-energy westerly winds that undulates over all our hemisphere with the exception of the low latitudes. In fact, when the jet does manage to reach into the tropics and encounters a hurricane, the hurricane's days, if not hours, are numbered because of massive wind shear. The top of the storm can be blown a hundred miles away from the bottom. Consequently, the same mechanism that causes extratropical cyclones is one that destroys hurricanes. You would think, then, if global warming were making extratropical storms stronger, there should be some concomitant weakening of hurricanes.

The jet stream is nature's way of dissipating the temperature difference between polar and tropical regions in the form of motion. The greater the temperature difference between the poles and the tropics, the stronger the jet, and, everything else being equal (dangerous words), the stronger extratropical storms can become. But the reverse is what should happen.

As noted in chapter 1, changes in atmospheric carbon dioxide result in a preferential warming of the coldest days, and of cold, dry air more than warm, moist air. Changing the greenhouse effect then must *reduce* the temperature contrast between the (warm, moist) tropical and (dry, cold) polar regions, which reduces the temperature difference that drives the jet stream. In turn, this should tame the power of extratropical cyclones.

Computer models nonetheless indicate that some rather small regions might see an increase in extratropical cyclones. The November 17, 2007, *Synthesis Report* of the IPCC is a 23-page document that attempts to summarize thousands of pages of the 2007 "Fourth Assessment Report" on climate change. It contains only two oblique references to increasing extratropical storms, talking about "increased erosion due to storminess" in Europe and "increases in the severity and frequency of storms" affecting coastal development in Australia and New Zealand. New Zealand is so far south that hurricanes are exceedingly weak and rare, so this reference *must* mean extratropical cyclones.

In the original 1,009-page science section of the "Fourth Assessment Report," there's one page, in the chapter on global climate projections (chapter 10), devoted to extratropical cyclones. It features a large number of citations and very dense prose. Some computer models increase their frequency, some decrease it. Others increase intensity, and still others decrease intensity. About the only thing that is agreed upon is something that doesn't require a computer model: if you preferentially reduce the amount of cold air with a changing greenhouse effect, the average track of these storms will shift slightly poleward. The notion of storm "tracks" itself is a little misleading, as low pressure systems can appear and travel to anywhere. "Track" just means that more storms tend to appear in some places rather than others.

In the IPCC report's next chapter, there are some projected changes, given with "low" confidence. They include a "decrease in the total number," and a "slight poleward shift of storm track" of extratropical cyclones, an "increased number of intense cyclones" and an "increased occurrence of high waves."

One reason for "low" confidence may have to do with the behavior of extratropical cyclones as the greenhouse effect has enhanced.

Here's the expanded (and very testable) statement from chapter 11 of the IPCC's 2007 science report:

> [Low confidence in] [i]ncreased number of intense cyclones and associated strong winds, particularly in winter over the North Atlantic, central Europe, and Southern Island of New Zealand.

What has happened as the planet warmed?

North Atlantic and European Cyclones

Christoph Matulla of Environment Canada and colleagues took a look at European storminess and published the results in the 2007 volume of the refereed journal *Climate Dynamics*.

Matulla et al. began by noting, "In the North–East Atlantic and the North Sea, a roughening storminess was perceived and public concern was raised in the early 1990s." Of course, in the early 1990s global warming hype shifted into high gear, especially in Europe.

Matulla et al. note that others have investigated trends in storminess in Europe over the time scale of 100 years, and on the basis of daily wind data, found no trend. However, long-term wind data "are characterized by spatial sparseness and inhomogeneities, caused by instrumentation changes, site moves and environmental changes." They state that this fact "highlights the importance of employing data that reach far back in time before any judgment about storminess can be made."

The scientists argue that "High wind speeds across Europe are generally associated with extratropical cyclones, which occur in North or North-Western Europe all year but in Central Europe almost entirely from November to February." Therefore, if we had evidence of the strength of the cyclones, we would have a way to detect if they have become more or less fierce in recent decades. Some of our longest weather records come from European locations. Many include barometric pressure—a direct measure of an extratropical cyclone's intensity—which is given by the difference in pressure between the outer and inner regions of a storm. The greater the difference, the stronger the wind.

Using long barometric records, they calculated daily winds back to 1875. Figure 5.1 shows the results. They found that northwestern European storminess starts at rather high levels in the 1880s, decreases to below-average conditions around 1930, and generally continues declining through the 1960s. From then until the mid-1990s, a pronounced rise occurs, and levels similar to those early in the century are reached. Since the mid-1990s, storminess is around average or below. This picture—a decline that lasts several decades followed by an increase from the 1960s to the 1990s and a return to calmer conditions recently, is found in Northern Europe, too. The increase, however, is far less pronounced. In Central Europe, storminess peaks around the turn of the 20th century, followed by a rapid decrease to 1960. That is followed by the familiar increase to the 1990s, with the most recent years showing a return to average conditions.

Figure 5.1
INDEX OF 99TH PERCENTILE OF DAILY WIND STRENGTH FOR EUROPE, 1875–2002

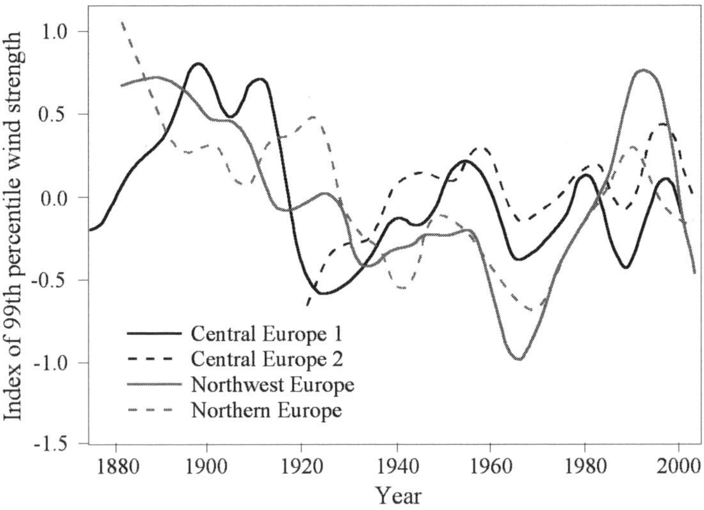

SOURCE: Adapted from Matulla et al. 2007.

They conclude that their work is in agreement with other studies in Europe showing "that storminess has not significantly changed over the past 200 years."

This study was predated by a much longer history over a much smaller geographic region. Writing in the 2004 edition of *Geophysical Research Letters*, Lars Barring and Hans Von Storch were very concerned about the public perception of increased storminess and global warming. They wrote:

> The public and ecosystems in storm-prone areas . . . are well adjusted to the continuous stream of passing windstorms. However, every now and then extreme windstorms cause severe damage. Together with the perspective of anthropogenic climate change, such extreme events create the perception that the storm climate would change; that the storms lately have become more violent, a trend that may continue

> into the future. The question is, of course, whether this perception is essentially caused by certain deeply rooted cultural notions about the relationship between man and nature, or whether such changes are real.

Again, they used barometric records, this time extending back more than 200 years for two locations in Sweden—Lund and Stockholm.

The barometer was invented by Torricelli in the mid-17th century, and some routine pressure observations can be found from the 18th century, although they typically are sporadic.

But the Lund and Stockholm data are of remarkable quality, with consistent readings as far back as the late 1700s. Station pressure readings were taken at least three times daily at Lund since 1780 and at Stockholm from 1820. Barring had examined these records for potential biases and inhomogeneities in prior publications and has developed a very long history of air pressure at these two sites.

In the current paper, the authors looked at three variables: the annual number of "storms" (station pressure less than 980 millibars [mb]; 28.94 inches of mercury on your home weather station); the annual number of observed pressure drops of more than 16 millibars (0.47 inches) in 12 hours; and extremes in the within-year distribution of 12-hour pressure changes. Each of those variables was then examined over the entire period of record to look for evidence of climate change.

A low pressure system of 980 mb is deep enough to generate winds capable of some damage. A change of 16 mb in a day requires a very active jet stream, which is one of the main sources of power for the common cyclone.

Figure 5.2 shows the long-term record of the annual number of observations of pressure below 980 mb. The records show very little change. The smoothed lines fitted through the data to better present long-term variations do show some minor undulations.

Of course, greenhouse gases have been increasing over this entire period (albeit only very slowly back in 1780), but the concentration of carbon dioxide shot up most rapidly since the mid-20th century. If you looked at records since only World War II, you could spot a teensy increase to the 1980s. But in the context of this more complete long-term record, it was equally stormy in the 1860s–1870s, when greenhouse-gas changes were virtually nil. Further, since 1990, the

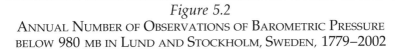

Figure 5.2
ANNUAL NUMBER OF OBSERVATIONS OF BAROMETRIC PRESSURE BELOW 980 MB IN LUND AND STOCKHOLM, SWEDEN, 1779–2002

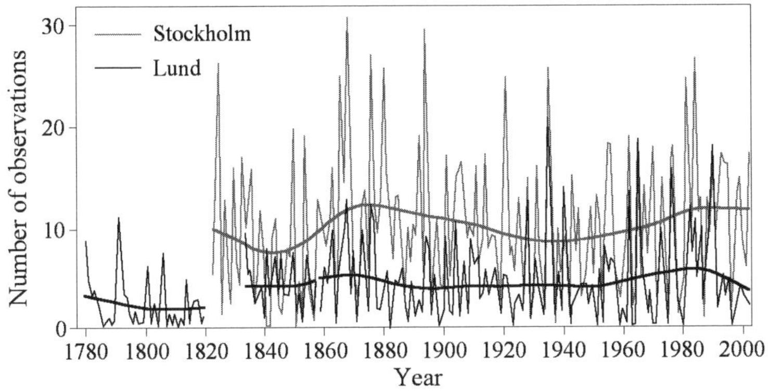

SOURCE: Adapted from Barring and Von Storch 2004.

pressure readings appear to have settled back to values near their long-term average.

The researchers' main conclusions?

1. "No significant robust long-term trends"
2. "The conspicuous increase in [the frequency of large 12-hour pressure drops] in Stockholm in the 1980s is evident but much less pronounced in the other storminess indices for Stockholm."
3. "The 1860s–70s was a period when the storminess indices showed general higher values of comparable magnitude as during the 1980s–90s. However . . . it is also clear that the indices have returned to close to their long-term average."
4. "The time series are remarkably stationary in their mean, with little variations on time scales of more than one or two decades."

They write: "[Our results] support the notion of an amplified storminess in the 1980s, but show no indication of a long-term robust change towards a more vigorous storm climate."

All climate records, particularly records prior to the 20th century, should be viewed with a healthy dose of skepticism. But those Scandinavian pressure histories are of remarkably high quality and simply show no evidence of unusual storminess as the planet warmed.

Nice Timing: *New York Times* Splashes Unrefereed Science during UN Global Warming Confab

On December 3, 2007, 15,000 climate bureaucrats, under the auspices of the UN, descended on Bali to discuss what is going to happen after the failed Kyoto Protocol on global warming expires in 2012. If everyone adhered to it, Kyoto would reduce global warming by 0.13°F (0.07°C) by 2050. The treaty was supposed to reduce net emissions of carbon dioxide by a bit over 5 percent. Instead, emissions rose approximately 5 percent ("approximately," because different sources, such the U.S. Department of Energy and the European Environmental Agency, among others, give slightly different figures).

Actually, the 2007 Bali confab was to discuss *what to discuss* next! Hey, isn't this what the Internet is for—electronically communicating unless travel is really necessary? Who cares about all the carbon dioxide emitted by the jets of the climate *nomenklatura*? It was winter in the Northern Hemisphere, and Bali is hot.

It hardly seemed an accident, when, on December 5, an article appeared in the *New York Times* extensively citing a nonrefereed study from a climate lobby called "Environment America" claiming that "extreme" precipitation has been increasing across the United States in the last half-century. Too bad that the sponsor, the Pew Charitable Trusts, didn't even get a refereed scientific article for their money (though they did hit the *Times* without one).

The *Times* quoted Environment America: "Across the United States, the number of severe rainfalls and heavy snows has grown significantly in the last half-century, with the greatest increases in New England and the Middle Atlantic region." Of course, the *Times* mentioned that that was just what was predicted to occur from global warming.

While most American farmers think more precipitation is a good thing, as evaporation exceeds normal rainfall most every summer, Environment America was quick to warn that "An increase in the frequency of storms delivering large amounts of rain or snow does not necessarily mean more water will be available" and that "[w]hile it may seem like a paradox, scientists expect that extreme downpours will be punctuated by longer periods of relative dryness, increasing the risk of drought."

Traditionally, the *Times* only writes about refereed science. Here they reported on none and ignored a paper that had appeared in

Geophysical Research Letters only the day before (by David Brommer, of the University of Alabama, and two coauthors), which concluded, with less rhetoric, that there has been a *very slight* increase in the frequency of the heaviest rainfalls in the United States. Indeed, none of this is news. A similar finding appeared by Michaels et al., in 2004 in the *International Journal of Climatology*

Environment America makes alarmist claims, tests the one claim it knows it can prove, does not discuss those findings in the proper context, chooses not to investigate claims that likely won't be supported by the data, and then throws in some factual errors for good measure.

Environment America asserts the following:

> In summary, scientists expect global warming to alter general precipitation patterns over the contiguous United States in four key ways:
>
> • Storms with extreme rates and amounts of rain or snowfall will become more frequent.
>
> • Summers will tend to be drier while winters will be wetter. Total precipitation will increase over most of the country but not in the Southwest.
>
> • The frequency of extreme events will increase much more than total precipitation.
>
> • Precipitation will become increasingly likely to fall as rain rather than snow— a simple consequence of increased temperatures. Paradoxically, the number of dry days will also increase, because intense downpours will punctuate longer intervals of relatively dry weather.

Those assertions are all testable with the precipitation data set compiled by Environment America, but the only one the group examined was the first one. And, given that annual total average precipitation has shown a general increase of about 10 percent in the United States over the past century or so (Figure 5.3, top), it was a pretty safe bet that the number of "extreme" precipitation events must be increasing as well.

Why a safe bet? Because, based on the nature of the distribution of precipitation events, there is a strong association between total precipitation and precipitation from heavy or extreme events. If you think about it, it's obvious. You can't very well get a lot more precipitation from a lot of light precipitation events—it takes 20

Figure 5.3
ANNUAL PRECIPITATION (TOP) AND ANNUAL AVERAGE FREQUENCY OF EXTREME PRECIPITATION EVENTS ACROSS THE UNITED STATES (BOTTOM), 1948–2006

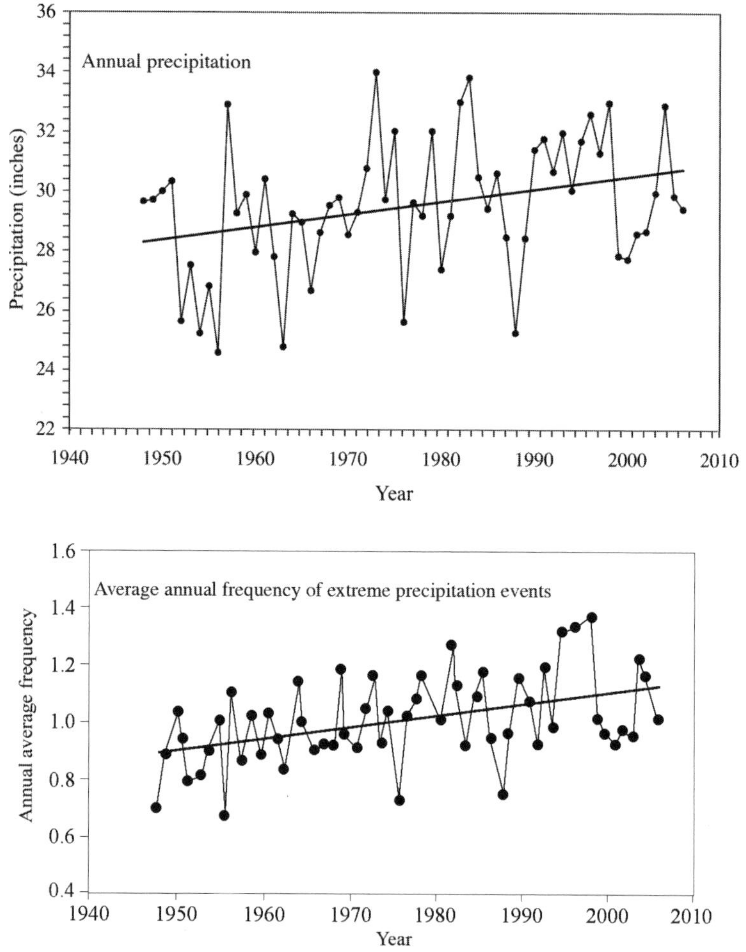

SOURCE:: National Climatic Data Center (top), http://www.ncdc.noaa.gov/oa/climate/research/cag3/na.html; Environment America (bottom) Environment America, http://www.environmentamerica.org/home/reports/report-archives/global-warming-solutions/global-warming-solutions/when-it-rains-it-pours-global-warming-and-the-rising-frequency-of-extreme-precipitation-in-the-united-states.

additional 0.10-inch events per year to contribute as much additional precipitation as one 2.00-inch event, or four 0.5-inch events. Twenty additional rain-days in a year is a large change while a few extra days with thunderstorms are hardly noticeable. By and large, the total annual precipitation that a location in the United States receives is highly dependent on the number of heavy or "extreme" rain days. This is true now, it was true 100 years ago, and it will be true in the future.

This becomes obvious when comparing the findings from Environment America for changes in extreme precipitation frequency (Figure 5.3, bottom) to a plot of the annual total precipitation averaged across the United States (figure 5.3, top).

Notice how closely Environment America's "average annual frequency of extreme rainstorms" tracks the observed average total annual precipitation across the United States. Both data sets show low values in the 1950s (when the United States was in a major drought), an increase from the 1950s to the early 1980s, low values in the late 1980s, high values in the mid-1990s, low values in the late 1990s to early 2000s and near-average in 2006, the last year in the study. Overall there is an upward trend from the drought conditions of the 1950s to the general wetness of the past few decades. That close correspondence between the frequency of extreme precipitation events and total average precipitation is what we wrote about in 2004 in *International Journal of Climatology*, and what is by and large the way things have to be. The higher the precipitation, the more of it comes via heavier events and vice versa. So if the climate is changing so that we get more precipitation, it virtually has to be the case that more of it will come in heavy or extreme events. *C'est la vie.*

So while Environment America's claim had to be true, it was a no-brainer.

The second claim, that summers will tend to be drier while winters will be wetter, could have easily been tested.

Alas, Environment America presents no seasonal data. Wonder why? Well, it is because it likely wouldn't have found any evidence for this assumed interseasonal behavior.

Figure 5.4 shows seasonally averaged U.S. precipitation data from 1948 to 2006 by season. Reality is virtually the opposite of Environment America's assertion: There is no trend in winter precipitation; spring, summer, and fall seasons all show slight increases.

Figure 5.4
U.S. PRECIPITATION BY SEASON, 1948–2006

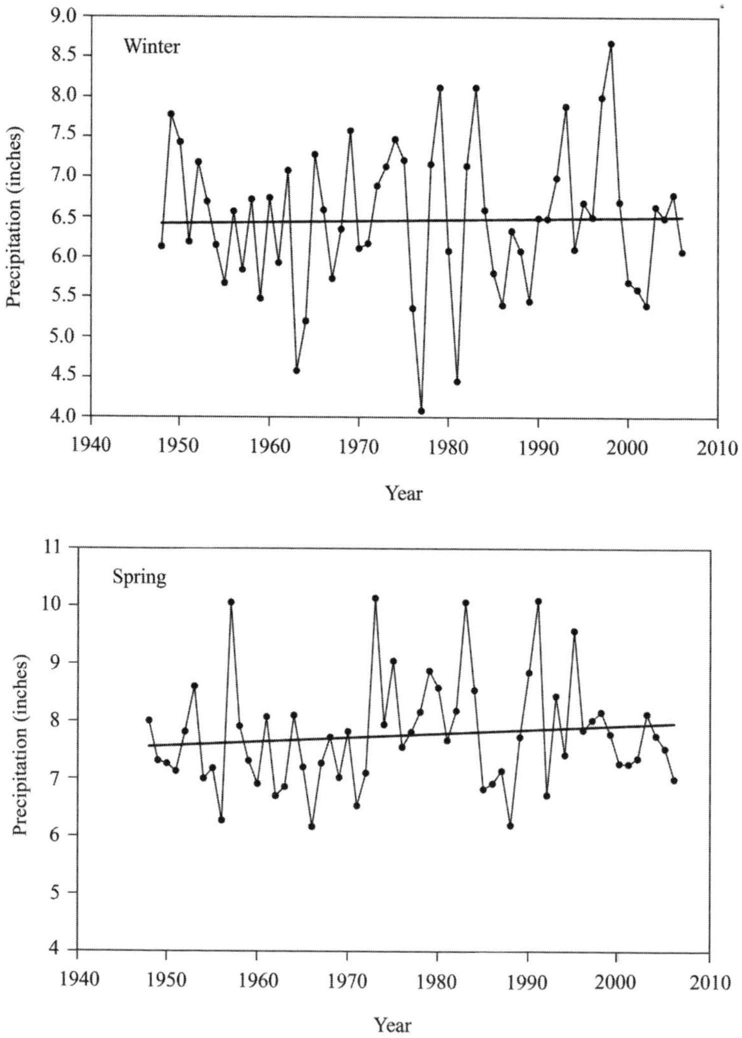

What about the third claim, that the "frequency of extreme events will increase much more than total precipitation?" Another easily

Figure 5.4 (continued)

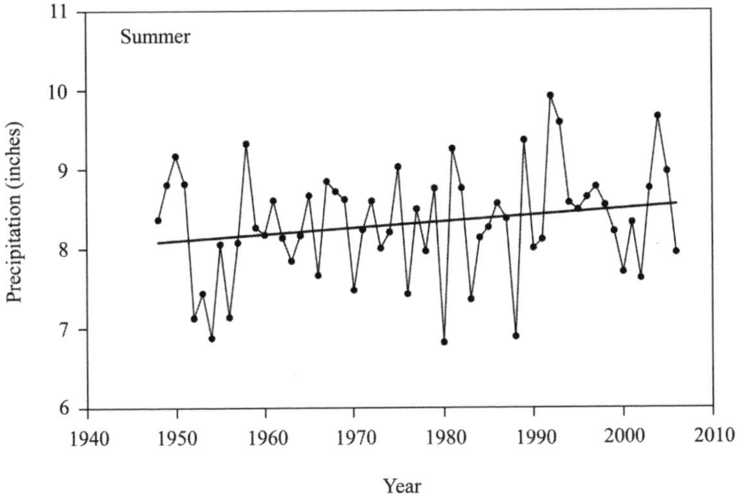

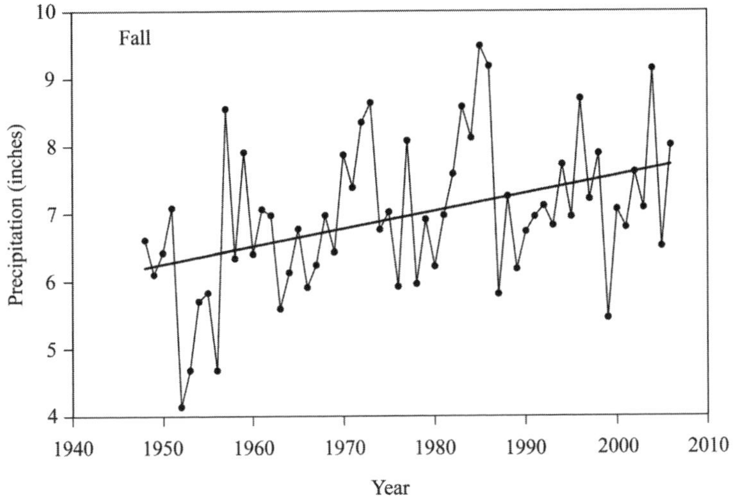

SOURCE: National Climatic Data Center 2007. http://www.ncdc.noaa.gov/oa/climate/research/cag3/na.html.

tested hypothesis given the group's precipitation data set, but no results are presented. Why not?

Figure 5.5 depicts what we published in our 2004 paper, showing the amount of precipitation that falls on the wettest day each year (i.e., "extreme" precipitation) averaged across the United States from 1910 to 2001. Notice that we indeed found that the wettest days of the year were getting wetter, or to put it another way, precipitation during extreme events was increasing. But contrary to the expectation of Environment America, the amount of precipitation falling on the wettest day as a percentage of the annual precipitation did not change at all over the same period (Figure 5.5). In other words, precipitation amounts in extreme events were not increasing more than total precipitation, or, to put it another way, the increase in extreme precipitation was not "disproportionate" when compared with the overall precipitation.

Half of Environment America's fourth claim, that "precipitation will become increasingly likely to fall as rain rather than snow—a simple consequence of increased temperatures"—is true (for the climate of the United States—not so in Antarctica) as demonstrated in a 1999 paper in *Journal of Geophysical Research* by Robert Davis et al. As to the second part of the claim, that "paradoxically, the number of dry days will also increase, because intense downpours will punctuate longer intervals of relatively dry weather," there is no indication that that is happening at all.

The people at Environment America could have tested this claim by simply tallying the number of dry days in the precipitation data set each year and seeing if the number was increasing. But, apparently they chose not to (or did so and didn't like the results), because they present no analysis of their own to support the claim. Instead, they attempt to use the scientific literature to do so. They fail miserably, mischaracterizing the results they cite and failing to cite other results that show that both the number of rainy days and the amount of soil moisture has been increasing across the United States.

On page 23 of its "When It Rains, It Pours" report, Environment America writes: "Since the 1970s in the contiguous United States, an apparently unusual increase in precipitation intensity has occurred. At the same time, the annual number of days with rain or snowfall has decreased."

Figure 5.5

AVERAGE AMOUNT OF PRECIPITATION (TOP) AND PERCENTAGE OF ANNUAL PRECIPITATION (BOTTOM) THAT FELL ON THE WETTEST DAY OF EACH YEAR ACROSS THE UNITED STATES, 1910–2001

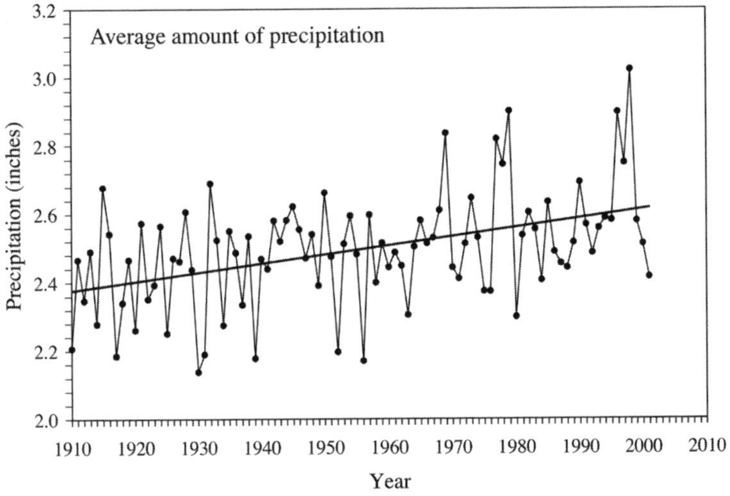

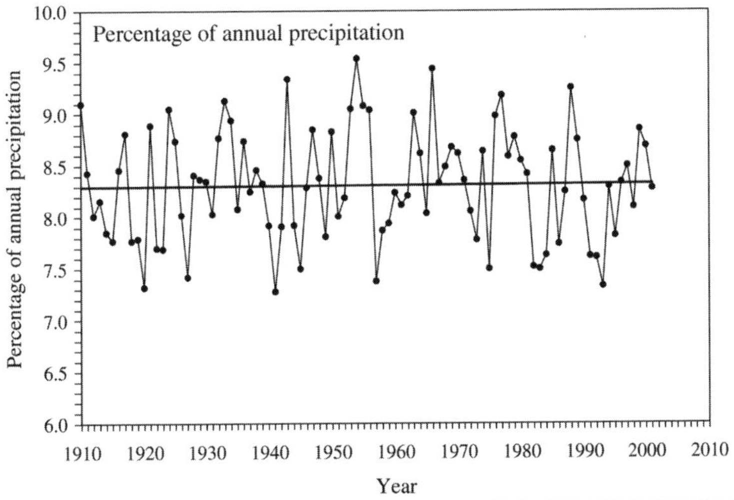

SOURCE: Adapted from Michaels et al. 2004.

We know total precipitation, in general, across the United States has been increasing, but have never heard that the number of precipitation days was decreasing. No reference was given for this statement. One reference that we know of that did examine the trend in precipitation days across the United States was Karl and Knight, so we checked that source. They concluded: "Clearly, the total annual increase in precipitation frequency [across the United States] of 6.3 days per century significantly contributes to the increase in precipitation." In other words, precipitation across the U.S. has been increasing, in part due to *increases in the number of days with precipitation*, the *opposite* of what Environment America reported. Another scientific journal article published in 2006 in *Geophysical Research Letters*, by University of Washington's Konstantos Andreadis and Dennis Lettenmaier, investigated trends in 20th century drought characteristics across the United States, and found: "Droughts have, for the most part, become shorter, less frequent, and cover a smaller portion of the country over the last century," completely the opposite of Environment America's claims.

So just where did Environment America get the idea that the number of days with rain was declining across the United States? From page 23 of their report:

> In 2002, Vladimir Semenov and Lennart Bengtsson at the Max Planck Institute for Meteorology in Germany compared actual observations of precipitation intensity with the results of two climate models over the contiguous United States during the 20th century. They found general agreement between the model and reality in terms of the trend toward more frequent extreme precipitation. They also observed that for the northeastern quadrant of the United States, the annual number of days with precipitation has been declining since the 1970s (simultaneously with an increase in the frequency of extreme downpours)—and the model generally reproduces the trend, albeit overestimating the absolute number of days with precipitation. [ref. 61]

Their reference 61 is this:

> V. A. Semenov and L. Bengtsson, "Secular Trends in Daily Precipitation Characteristics: Greenhouse Gas Simulation with a Coupled AOGCM [Atmosphere-Ocean General Circulation Model]," *Climate Dynamics* 19: 123–40, 2002.

Did Semenov and Bengstsson report that "for the northeastern quadrant of the United States, the annual number of days with precipitation has been declining since the 1970s"? They wrote no such thing. In fact, what Environment America interpreted and reported as observed changes across the northeastern United States were actually climate model simulations of the precipitation characteristics there. So Environment America's lone piece of support for its contention that there has been a decline in rainy days across the United States is not based upon *observations* (which show an increase in rainy days), but instead upon a *climate model simulation* that was wrong.

All the claims that Environment America made about precipitation across the United States were testable. Environment America chose (or only reported on) the few that it knew had to be correct simply on the basis of the general characteristic of the weather and the weather trends over the United States during the past 50 years (that is, the more rain you get, the more rain comes from "extreme" events, and the warmer it gets, the less precipitation falls as snow). Its other claims were dead wrong.

So, what we are left with is nothing but a basic climatology lesson from Environment America, with a soupçon of untested (but easily tested) alarmist assertions that turn out to be false.

What's strange here is how this unrefereed, loosey-goosey study got the attention of the *New York Times* during an important UN conference on climate change, when the real, hard science was out there for all to see.

Another View of Extreme Rainfall

The notion that human-induced climate change is making more extreme weather is everywhere. A search for "extreme weather + climate change" will get you 1.5 million hits on Google. If it makes for that much Internet traffic, there *must* be a connection, right?

The Illinois State Water Survey's Ken Kunkel presented some new results and a summary of some recent papers on extreme weather in the continental United States at the Climate Specialty Group's plenary session at the Association of American Geographers annual conference in Chicago in 2006. His analysis shed some new light on long-term extremes in heavy precipitation based on new data.

Figure 5.6
RELATIVE FREQUENCY OF CO-OP STATIONS EXPERIENCING A SINGLE-DAY RAIN TOTAL NORMALLY OCCURRING ONCE IN 20 YEARS, 1895–2000, 7-YEAR RUNNING AVERAGE

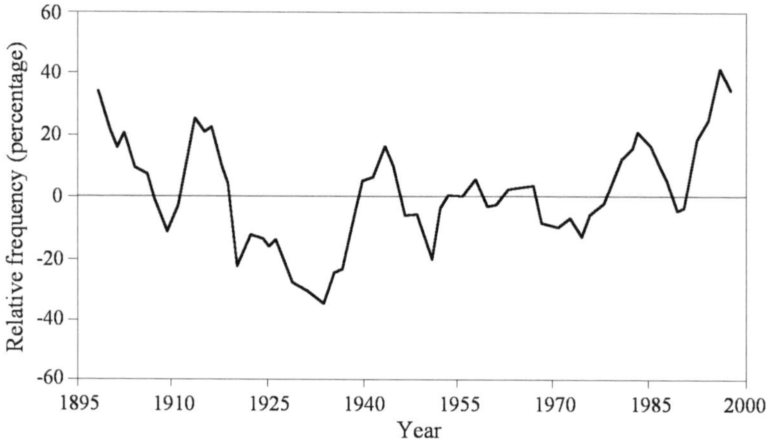

SOURCE: Adapted from Kunkel 2006.

Since the late 1970s, records from the co-op network (see chapter 2) were available in computer-readable form only back to 1948. Recent interest in climate change spurred an effort to digitize the earlier paper records, adding many new stations and much more data to the existing network. Now, the computer-readable precipitation database extends back to 1890, and the number of stations has more than doubled.

Precipitation tends to be pretty spotty, which means that a lot of new data dramatically increase the number of places that could experience, say, a single-day rainfall that would normally occur only once in 20 years. Kunkel calls this number the Extreme Precipitation Index. We show his values for the number of stations receiving a once-in-20-years single-day rain total in each year back to 1900 (Figure 5.6). Positive values of the Extreme Precipitation Index indicate above average occurrence of extreme events, whereas negative values mean less frequent extremes. The data since 1950 show a clear positive trend in the index, which fits nicely with all the scare stories.

But inclusion of the pre-1950 data paints a much different picture. The frequency of extreme rainfall in the late 1890s was at least comparable to that in our current climate. Kunkel did some statistical tests demonstrating that the most recent period (1983–2004) is not statistically different from the earliest period (1895–1916). The bottom line? The assertion that U.S. rainfall is clearly getting more extreme because of global warming can't be supported if the frequency of extreme rain was as great 100 (colder) years ago as it is now.

The Fire This Time

"Fire" and "drought," at first blush, should go hand-in-glove. But, like a lot of other things about global warming, what people think should be happening isn't necessarily what *is* happening.

Take the huge Southern California conflagration of October 2007. Senate Majority Leader Harry Reid (D-NV) blamed it on global warming, saying in Washington's authoritative political newspaper, *The Hill*, on October 24, "One reason we have the fires in California is global warming."

Global warming is a great issue, because it affords such easy opportunities for climate scientists to test glib hypotheses made by politicians who often have no training whatsoever in climate science. Reid's statement is particularly easy to test.

Don't blame California wildfires on summer's heat. By the end of each and every summer (to paraphrase the song, "it never rains in California" in that season), Southern California is drier than the world's best martini.

California's big wildfires are, ironically, caused by excessive winter rains. Normally, the Southern California region that blew up in 2007 averages about a foot from December through March, the local rainy season. That moisture turns Southern California green in the winter and early spring. Owing to the fact that just about every day after the rainy season is warm and sunny, it's only a matter of a month or two before the surface dries out to the point that there's not enough water to support additional plant growth, and Southern California dries up.

The distribution of Southern California rainfall from year to year is a bit unusual. The vast majority of the years have below-normal precipitation—about four or so inches below average. But in the few years that are above average, when it rains, it pours, with rainfall often 100 percent (one foot or more) above the mean.

Some of the very wet years are caused by El Niño—which you'll recall is a reversal of winds every few years over the tropical Pacific Ocean that has been going on for millenia (see chapter 1). People such as Senator Reid (and Vice President Gore) will cite computer models predicting that El Niños should become stronger or more frequent with global warming, but there are other models that show they won't change or that they might lessen in frequency. The last big El Niño was in 1998, and we are *way* overdue for a strong one (which will probably reset the surface temperature record, which also dates back to 1998).

So, some computer models say global warming means more El Niños, which means more vegetation, which means more fire.

When things get very wet, there's plenty more time for the soil to remain moist, producing a much longer growing season in the hills where suburbs and very expensive homes are proliferating. The problem is that these rooms-with-a-view also are houses-with-a-risk; that is, they're in the path of wildfires.

If Senator Reid is right, and the catena from global warming to wildfires is mediated by more vegetation, then rainfall, or the frequency of rainy years in Southern California must be increasing as the planet warms. Like so many others, Reid's is a very testable hypothesis.

Figure 5.7 depicts the total December–March precipitation for the California South Coast Drainage Climatological Division from 1895 through 2007.

Remember, most of the years are below the long-term average of about 12 inches, but the relatively few that are above the mean are often *way* above it. If global warming is causing the increase in Southern California wildfires, then the frequency of very wet years has to be increasing in a significant fashion.

Obviously, it is not. In fact, the biggest agglomeration of far-above-normal years was a 12-year period beginning in 1905.

Ironically, those rains were concurrent with some of the massive westward migration of U.S. population, in an era when both California and Arizona were touted as green paradises, which they were, thanks to all that unusually lush vegetation. Sure, there had to have been enhanced wildfires back then, but very few people lived within their reach.

Figure 5.7

DECEMBER–MARCH RAINFALL FOR THE CALIFORNIA SOUTH COAST DRAINAGE CLIMATOLOGICAL DIVISION, 1895–2007

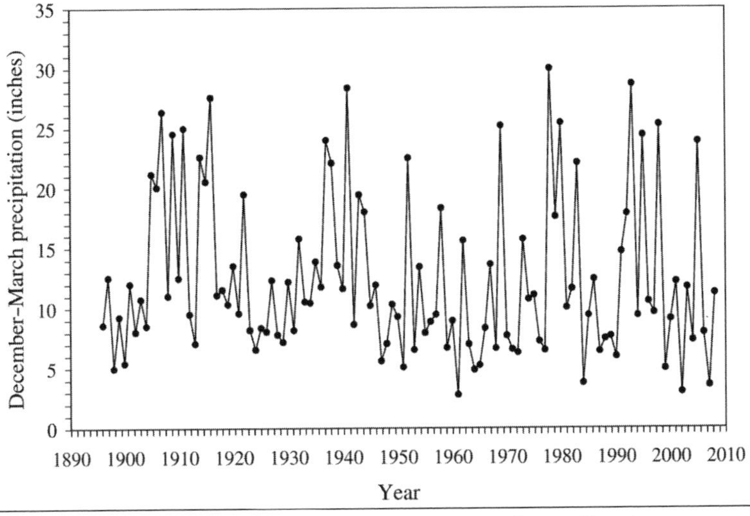

SOURCE: National Climatic Data Center 2007. http://www7.ncdc.noaa.gov/CDO/CDODivisionalSelect.jsp.

After a very wet year (2005 being the last big one), it's only a matter of time before a thousand or more homes get torched in the hills around Los Angeles.

But don't blame the 2007 Southern California conflagration on global warming: There is no trend whatsoever in the frequency of heavy-rainfall years that would promote wildfires. And our public officials could do us all a service by not making statements like Reid's to *The Hill*, which almost all U.S. Senators and Congressmen read and therefore believe.

Southwestern Drought: The Long-Term Perspective

Some 38,000,000 Californians largely (but not completely) depend upon two water sources: winter snowpack in the Sierra Nevada and the Colorado River, which forms the eastern boundary of the state with Arizona.

So much water is drawn out of the Colorado River by Californians and everyone else that, by the time it reaches its mouth at San Luis

Rio Colorado on the Gulf of California, it's barely a trickle. Most streams increase in volume as they approach the sea, but not the Colorado, thanks to the huge amount of water withdrawn by the citizens of the burgeoning Pacific Southwest. Consequently, it's a guaranteed front-page headline when anyone says that global warming will increase drought frequency, reducing the Colorado's flow. Such assertions, alas, like Senator Reid's linking of global warming to fires, are eminently testable.

California has apparently warmed up about 2°F (1.1°C) since 1950, and Arizona has warmed 4°F (2.2°C). We use the word "apparently" because former California State Climatologist James Goodridge detected a strong urban warming bias in California temperatures, which he published in the *Atmospheric Environment* in 1992. Given that Arizona has recently experienced major urbanization, the same thing is likely there.

David Meko, at the University of Arizona, and colleagues studied the relationship between tree rings and streamflow. Desert trees are very responsive to rainfall, so the correlation between the width of an annual ring (the yearly growth increment) and precipitation (and therefore streamflow) is very high.

There are also excellent records of Colorado River streamflow back to 1906, which gave Meko almost 100 years of data to compare with the tree rings. Some of the trees out there are very old, and Meko was able take the known relationship between streamflow and tree rings all the way back to the year 762.

Apparently there is nothing new under the California sun. In Meko's words:

> The most extreme low-frequency feature of the new reconstruction, covering A.D. 762–2005, is a hydrologic drought in the mid-1100s. The drought is characterized by a decrease of more than 15 percent in mean annual flow averaged over 25 years, and by the absence of high annual flows over a longer period of about six decades.

Figure 5.8 shows Meko's record of observed and tree-ring-constructed streamflow. It is obvious that there is simply nothing unusual in this record that is concurrent with the planetary warming that began around 1975. For what it's worth, the average flow for the period of historical record (1904–2005) is generally larger than the estimated flow from 762 to 1905.

Figure 5.8

Observed Colorado River Streamflow, 1905–2005, and Streamflow Reconstructed by Use of Tree Rings, 762–2005

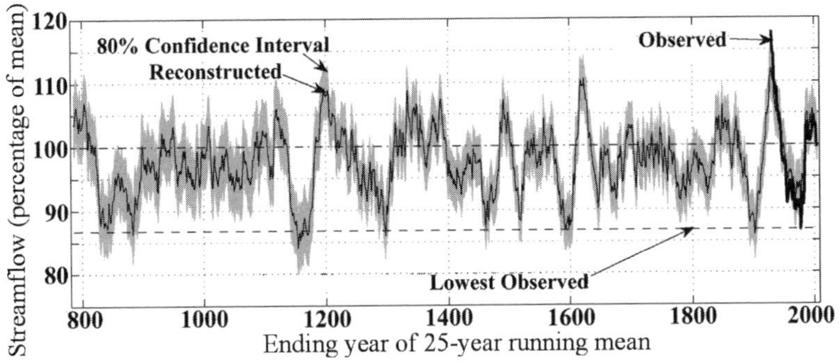

Source: Adapted from Meko et al. 2007.

Anyone suggesting that recent droughts are somehow the result of emissions of greenhouse gases is overlooking a tremendous piece of evidence suggesting otherwise. Droughts are a natural part of the climate of the Pacific Southwest—they have been around a long time, and they are not going away anytime soon. Droughts impacted the region in warm periods of the past and cold periods of the past, and will return whether the future is warmer or colder.

Global and Local Wildfires

What could be simpler? Global warming without a compensating increase in precipitation will make much of the world drier. Drier vegetation is more subject to immolation. There should even be a positive feedback: After all, vegetation burns largely to carbon dioxide and water, which should contribute additional warming. Couple that with the fact that most burnt areas are black (black surfaces absorb more solar radiation than white ones, which is why so much traditional Middle Eastern clothing is white), and . . . even more warming!

Back in the big El Niño of 1998, Al Gore traveled to Florida, where wildfires were rampant, and proclaimed that this was the world of the future. In other words, as the planet warms, there's going to be more fire.

Yet another eminently testable hypothesis made by a politician about global warming.

A 2007 article in *Global Change Biology* did precisely that. In the paper titled "Global Spatial Patterns and Temporal Trends of Burned Area between 1981 and 2000 Using NOAA-NASA Pathfinder," David Riaño, from University of California-Davis, and several colleagues examined 20 years of global satellite data. Conclusion? "The total annual burned area has not increased in the last 20 years."

Well, certainly the temperature has, so the chain of causation from warming to drying to burning to more warming to more drying to more burning mustn't be so simple!

Riaño et al. made the case that determining the burn area and/or biomass consumed annually would be critical in understanding various dimensions of the global ecosystem, but that to date, the data have been collected annually using highly irregular criteria from country to country. The UN Food and Agriculture Organization has tried to collect the statistics, but its data are notoriously suspicious in terms of accuracy. Riaño et al. argued:

> A single remote sensing data source can provide globally coherent multitemporal spatial information, not only from the visible part of the spectrum but also reflected solar infrared, which can be used to obtain consistent environmental monitoring at the global level.

Riaño et al. collected the 8-km-resolution global satellite-based Advanced Very High-Resolution Radiometer data from July 1981 to December 2000. They developed a computer algorithm that could spot pixels (the spatial units returned by satellites) in which there were recent fires, total them, and determine percentage of burned area for any defined region of the world. The global and regional data were ultimately assembled on a monthly basis.

In analyzing trends in the burn area data, his team found: "The global trend statistics in the total number of pixels burned in any month or annually were not significantly different from zero. . . . Therefore, no significant upward or downward global trend was found in the burned area data."

The research team then did a mathematical analysis, called cluster analysis, to identify areas with similar fire histories, and they reported that cluster group 1 did show a significant increase in burned area. This area included the western United States, where wildfires have received a tremendous amount of press recently. But they also found a significant *decrease* in the burn area that included Central America and much of Southeast Asia. When viewed globally (a good idea when looking at *global* warming), Riaño et al. once again reported, "There was also no significant upward or downward global trend in the burned area for any individual month."

What an inconvenient truth! The study period, 1981 to 2000, was a two-decade time period when greenhouse gases increased substantially in the global atmosphere, when the earth warmed, and when the media breathlessly covered every fire from Indonesia to Australia to the western United States. Many scientists are quoted in the nearly *two million* websites on wildfires claiming that the increase in fire activity is well under way.

Riaño et al. developed the first truly global data set on burn area, conducted an analysis that is uniform across the globe, and found no trend in the global burn area data. So much for getting your science off the Internet!

Riaño et al.'s study was preceded as is a more local analysis but one that was cleverly extended to long before the 1980 dawn of the fire-sensing satellite.

A 2006 article in *Journal of Geophysical Research* by M. P. Girardin and colleagues showed that the burned area from forest fires in Ontario, Canada, has increased since 1970. However, the increase in fires pales in comparison with what was observed in the more distant past.

At the outset, the authors note: "It is also possible that the provincial fire statistics (number and size of fires) were underreported prior to the 1960s." Girardin et al. were interested in extending data on the burned area of Ontario going back more than a hundred years, and they used tree rings to estimate burned areas of the past. Not surprisingly, forest fires produce a recognizable signal in the annual tree rings, and the Canadian team used sophisticated statistical wizardry to go from tree ring patterns to total area burned in a given year. They tested their statistical model from 1917 to 1981, and once satisfied that the model was working with reasonable accuracy, they used tree rings to extend burned area estimates back to 1781.

Figure 5.9 shows the results. According to the authors, "Episodes of succeeding years of large area burned were estimated approximately at 1790–94, 1803–07, 1818–22, 1838–41, 1906–12, 1920–21,

Figure 5.9
BURNED AREA IN ONTARIO, CANADA: RECONSTRUCTED, 1782–1981, AND OBSERVED, 1917–2003 (TOP); AND 10-YEAR SMOOTHED DATA, 1782–1980 (BOTTOM)

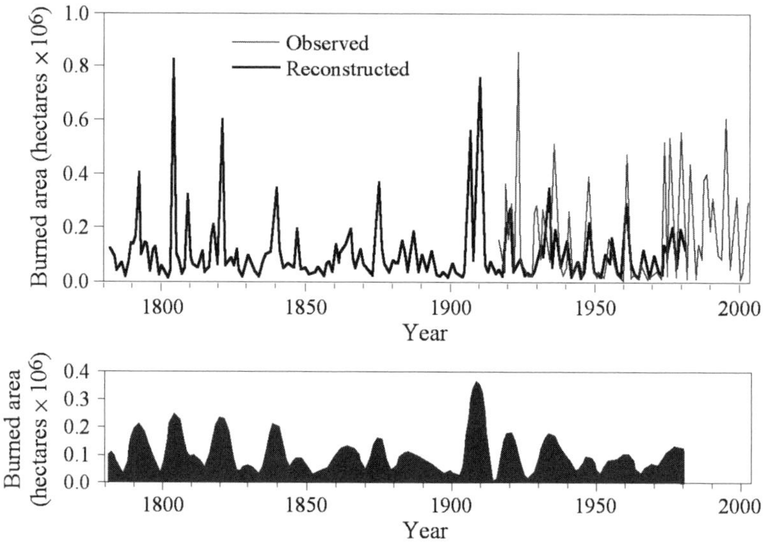

SOURCE: Girardin et al. 2006.

NOTE: Thin line in top figure represents observed rather than modeled data (1917–2003).
1 hectare = 2.47 acres.

and 1933–36.'' Furthermore,''The tree ring model revealed the year 1804 as the year of most extreme area burned.'' In addition, ''The sliding window analysis showed higher mean area burned values prior to 1840 and through the 1860s–80s and the 1910s–20s. Mean values during the mid-20th century were the lowest in the record.''

They concluded: ''The reconstruction indicated that the most recent increase in area burned in the province of Ontario (circa 1970–81) was preceded by the period of lowest fire activity ever estimated for the past 200 years (1940s–60s).'' Finally, they wrote, ''[W]hile in recent decades (circa 1970–81) area burned has increased, it remained below the level recorded prior to 1850 and particularly below levels recorded in the 1910s and 1920s.'' Figure 5.9 shows that the biggest burn periods all occurred prior to World War II.

What about an increase in fires in western North America?

Again, looking at the multicentury record, and the propensity for forest fires to leave their traces in tree rings, it appears that fire frequency for the last 500 years has been fundamentally the result of natural ocean climate cycles, and not global warming.

That is the conclusion reached by researchers from Colorado, Arizona, Montana, and Argentina in a 2007 study published in the *Proceedings of the National Academy of Sciences*. Lead author Thomas Kitzberger, from Argentina's Universidad Nacional del Comahue, and his coauthors used extensive tree-ring analyses to determine the underlying causes of these widespread and episodic wildfires.

The researchers examined more than 33,000 fire events from nearly 5,000 fire-scarred trees, primarily ponderosa pine and Douglas fir, throughout the western portion of North America. The exact calendar year of the fire event was obtained by noting the specific year of the tree ring record in which a fire scar occurred. The researchers investigated 238 sites throughout the western half of North America. This is an extensive study!

The scientists then linked the fire frequencies with measures of drought and corresponding sea-surface temperature. They concentrated on three specific oceanic climate phenomena: First, they obtained records of the sea-surface temperatures in the central equatorial Pacific in order to study the effects of El Niño on forest fires.

Another strong, but longer-term, ocean climate phenomenon in the Northern Pacific is called the Pacific Decadal Oscillation (PDO). The PDO is a pattern of temperature variation that changes very slowly—in a multi-decadal fashion—compared to the one- or two-year El Niños. The PDO changes whether or not there is global warming.

The third climate phenomenon the researchers related to fire frequency is a pattern of temperature variation in the north Atlantic known as the Atlantic Multidecadal Oscillation (AMO), which operates on even longer time frames, with periods of up to 60 to 80 years.

The researchers noted that there had, until then, been no comprehensive study linking the overall patterns of these oceanic oscillations with the overall patterns and frequency of fires in the western United States. Consequently, they took this massive database of 33,000 fire events derived from tree-ring records across the entire western North America and compared it with those ocean climate phenomena.

Not surprisingly, Kitzberger and colleagues found that, since the year 1550, fire events and droughts have been linked—the occurrence of drought and wildfire in the West have coincided. More important, they found those occurrences have been strongly tied to variations in the three ocean-climate phenomena—El Niño, the PDO, and the AMO. They found that the magnitude of the two Pacific climate phenomena, El Niño and the PDO, were crucial to determining the frequency of fires in subregions across the West. But given the recent spate of massive wildfires across that region, they were most interested in whether widespread fires—extending across large areas of the West—were linked to those oceanic climate variations.

They found that the ocean temperature variations associated with the AMO were the dominant factor. In particular, they found when the AMO is acting like it has since 1995 (also spawning some terrific hurricane seasons), there's a propensity for huge western conflagrations. They wrote: "The key issue is that the Atlantic Multi-Decadal Oscillation persists on time scales of 60 to 80 years, compared to just one year or a few years for El Niño."

So does their study have any implications for the future? Can we tell something about upcoming widespread fire occurrence in the western portion of North America based on this extensive tree-ring analysis? The answer is yes—and the implications aren't too good. Unfortunately, given the very long-term nature of this Atlantic ocean-climate phenomenon, we are likely to be in the AMO warm phase for quite some time to come. That suggests we will likely continue to see more and more massive western fires, global warming or not.

Further, it leads to the conclusion that big hurricane seasons and wildfires are correlated with the AMO. How hard will it be for anyone to *not* say the world is going to heck in a hurricane because of global warming, even though that's not the cause?

6. Climate of Death and the Death of Our Climate

Who can forget the massive European heat wave of 2003? Certainly not cyberspace, where Googling "global warming + mortality" will get you 723,000 hits. Although accurate numbers are virtually impossible to come by in such a situation, it appears that 15,000 people in France died from heat-related causes, and there were 35,000 total deaths in Europe.

Needless to say, this event was probably the single incident most responsible for the remarkable perseveration on global warming that afflicts the Continent. It certainly prompts two questions: (1) Was it climatologically unprecedented? and (2) What can be expected as the planet warms?

The first seems like a hands-down yes. After all, no heat wave in recorded history killed so many French. But a closer inspection reveals a much more complicated picture.

A 2006 article in *Geophysical Research Letters* dares to ask the question: "Was the 2003 European summer heat wave unusual in a global context?"

T. N. Chase and several colleagues from Colorado and France begin their study by noting:

> The European heat wave of summer 2003 has received considerable attention, both because of a potential link to larger-scale warming patterns (e.g., "global warming"), and the large loss of life. Several studies find that this regional heat wave was quite unique and it has been suggested that such an extreme event could be accounted for only by a shift of statistical regime to one with higher variance [i.e., a "changed" climate].

The argument is that climate change caused by the buildup of greenhouse gases increases temperature variability, and this increased variability makes heat waves like the one in 2003 more likely. Chase et al. decided to test this hypothesis.

Recall the discussion in chapter 1 about how changing the atmosphere's greenhouse effect preferentially warms the high-latitude land areas (despite the unexplained lack of warming in Antarctica), which reduces the temperature contrast between the North Pole and the equator, which reduces the strength of the jet stream and therefore reduces temperature variability. Perhaps a prolonged heat wave—where temperatures remain high from day to day—could be interpreted as such a reduction in variability.

The Chase team collected data on the temperature of the atmosphere from throughout the world for the surface to the midatmosphere for the period 1979–2003. For each month, they computed the mean and standard deviation of the temperature, thereby allowing them to map temperature anomalies in terms of standard deviations above or below the mean.

The standard deviation is a measure of the "spread" of data around the mean value. Think of it as variability. There's an average temperature for, say, the 4th of July at a given location in the United States. That average is calculated by taking all of the July 4 readings, adding them up, and dividing by the number of observations. Some days are warmer than the average, some are colder. Same for, say, New Year's Day.

But the variability of 4th of July temperatures from year to year will be less than that for New Year's. Why? In winter, the nation is subject to invasions of frigid arctic air, as well as milder conditions in its absence. In summer, the air masses over the country are of much more uniform (and hot) temperature. So the standard deviation of 4th of July temperatures will be smaller.

Roughly two-thirds of all observations of temperature at a given location and date are within one standard deviation of the mean. The probability of being two standard deviations above the mean is 0.05 (1 in 20), and three is 0.003, or 3 in 1,000.

Technically, Chase et al. used something a little different from temperature. Instead, they measured the distance from sea level at which a weather balloon would find one-half the atmosphere underneath it. The greater the height to which the balloon must ascend, the warmer the air is beneath it. Under cold conditions, the atmosphere is more dense, and the balloon doesn't have to ascend as far in order to be above half the atmosphere.

This distance is known as "thickness" to meteorologists, and it is directly proportional to surface temperatures absent any local factors such as lakes or forests or cities. Consequently, it gives a truer picture of the atmospheric temperature than one gets from most weather stations.

As seen in Figure 6.1 (see insert), for June, July, and August of 2003, Europe was definitely ground zero for what is an extreme, extreme event. Note that far more than half the planet is portrayed in green and blue tones, indicating normal to below-normal temperatures. Europe was simply located in the wrong place at the wrong time, and the heat wave was anything but global in nature.

The three-standard-deviation anomaly over Europe has a statistical probability of only 1 in 333. Given that there are a lot of places on the planet, and four seasonal slices to examine, it's just not that odd that such an anomaly shows up *somewhere*. But this time it happened to be at the epicenter of global warming fever (in the midst of a climatically moderate summer around the planet).

Chase et al. analyzed the thickness anomalies for all parts of the globe for the 25-year study period and concluded: "Extreme warm anomalies equally, or more, unusual than the 2003 heat wave occur regularly." They can occur at any time of the year. They will rarely appear in summer, directly over Europe, where many residents eschew(ed—see later text) air-conditioning. They also note: "Extreme cold anomalies also occur regularly and can exceed the magnitude of the 2003 warm anomaly in terms of the value of SD [standard deviation]." Of course, it's pretty tough to sell the idea that global warming is causing cold anomalies, so cold anomalies are not nearly so newsworthy.

It should come as no surprise that Chase and company noted that warm years tend to have heat waves and cold years have cold waves. But their next two conclusions are more interesting. They found:

> Natural variability in the form of El Niño and volcanism appears of much greater importance than any general warming trend in causing extreme regional temperature anomalies as regional extremes during 1998 [a year with a huge El Niño] in particular were larger than the anomalies seen in summer 2003 both in area affected and [standard deviation] extremes exceeded.

In summer 2003, 2 percent of the planet experienced thickness-temperature anomalies above two standard deviations. In the big El Niño year of 1998, that figure was nearly 30 percent. Three standard deviations above normal covered 5 percent of the planet for the entire year of 1998, while the *nowhere* on the planet exceeded this criterion in the year 2003.

Chase et al. also examined the trends in the data over the 25 years and reported: "Analyses do not provide strong support for the idea that regional heat or cold waves are significantly increasing or decreasing with time during the period considered here (1979–2003)." In other words, heat waves like the one in Europe in 2003 can and will occur by chance even if temperature does not rise or the variability of temperature does not change.

There is no question that the heat wave of 2003 was a natural disaster in Europe with a substantial loss of human life. Europe was not prepared for an event that, from a purely statistical view point, was inevitable, with or without global warming.

In 2006, another article appeared in the *International Journal of Biometeorology* that put the 2003 disaster in perspective. Mohamed Laaidi and two coauthors, from the Medical University at Dijon, France, examined daily temperature and mortality data from 1991–95 for six "departments" (a.k.a., states or counties) located in urban, oceanic, interior, mountain, and two different Mediterranean settings (Figure 6.2). They broke the data into three age groups including less than 1 year old, 1 to 64 years old, and greater than 64 years old. They also divided the data by sex and by major causes of death including respiratory disease, cerebrovascular disease or stroke, heart disease, and other diseases of the circulatory system. Murders and accidents were excluded.

The Laaidi et al. team found that for the whole population

> As expected, temperature and daily deaths exhibited a marked temporal pattern. For all the departments investigated, mean daily counts of deaths showed an asymmetrical V-like or U-like pattern with higher mortality rates at the time of the lowest temperatures experienced in the area than at the time of the highest temperatures.

The data also clearly showed that people adjust to their environments. Individuals living in cold regions experience more mortality in warm temperatures, and those from warm areas are more susceptible to cold ones. There is also a range in temperature, called the thermal optimum, in which mortality is low; the authors noted:

Figure 6.2
Six Study Areas in France with Different Climates

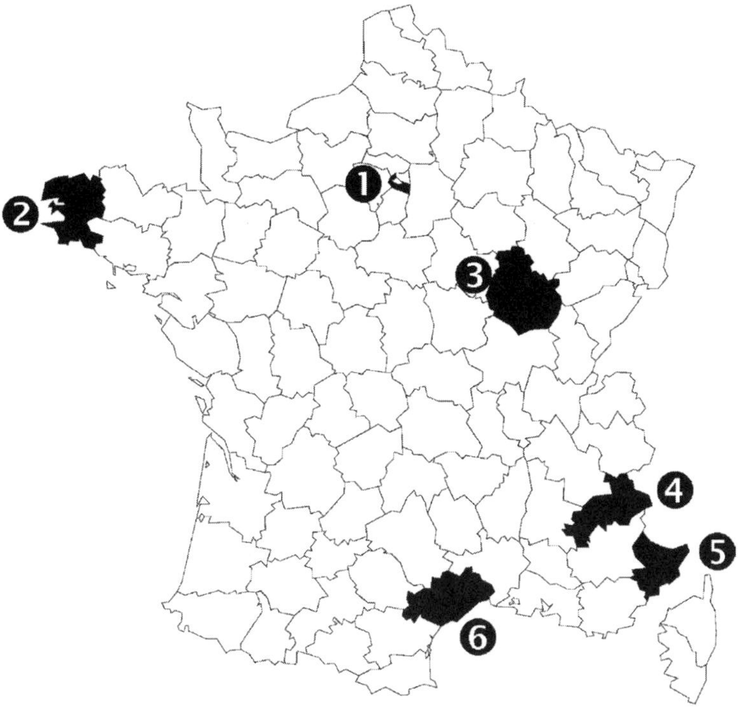

Source: Laaidi et al. 2006.

Note: 1 = Seine-Saint-Denis; 2 = Finistère; 3 = Côte d'Or; 4 = Hautes-Alps; 5 = Alpes-Maritimes; and 6 = Hérault.

> The level of the thermal optimum rises in line with the warmer climatic conditions of each department. The thermal optimum is greater in Paris, probably due to the urban heat island, than in the Hérault, which is situated in the extreme south of France in a Mediterranean climate.

In other words, here's the shocking news: People adjust to the climate in which they reside. In *Meltdown*, one of us (Michaels) cited work he had done with Robert Davis at the University of Virginia in which they found that heat-related mortality *declined* as cities get warmer, which cities do with or without global warming. The same phenomenon was seen by Laaidi et al., except they added in the adjustment for cold climates, showing less mortality there from cold waves than occurs when temperatures fall dramatically in warm climates.

Concerning any temperature rise for any reason, Laaidi et al. found: "For both men and women mortality was higher at low temperatures, suggesting a lesser ability to adapt to the cold." On the basis of another related study, they state, "In England and Wales, the higher temperatures predicted for 2050 might result in nearly 9,000 fewer winter deaths each year." Laaidi et al. conclude: "Our findings give grounds for confidence in the near future: the relatively moderate (2°C) [3.6°F] warming predicted to occur in the next half century would not increase annual mortality rates."

Computer models for carbon dioxide–induced global warming consistently predict more warming in winter in midlatitude locations such as France and less warming in the summer. The Laaidi et al. study shows that the greater threat of human mortality lies in the cold end of the thermal spectrum rather than the warm end. Higher temperatures in the winter would certainly decrease mortality, and we could conclude from this and other studies that in terms of temperature-related mortality, global warming would save lives—a message not well conveyed in the hundreds of thousands of websites on the subject.

Death vs. Life with Global Warming

The best way to make headlines in the global warming game is to generate scary scenarios about how many people are going to die.

It's little surprise then that a "Review" article by Jonathan Patz of the University of Wisconsin–Madison and colleagues that appeared in *Nature* in November 2005 caught tremendous attention. It focused on global warming and death.

Patz et al. began with the 2003 heat wave in Europe. As demonstrated earlier in this chapter, it was hardly unusual in the statistical sense and was more an accident of geography than anything else. (Note that the Chase et al. study appeared after Patz et al.'s review).

A substantial (and not discussed) cause of a large number of fatalities was cultural. The month-long August vacation is a cherished European tradition. It's not unusual for many countries to effectively shut down while the epicenter of the population shifts southward to Mediterranean beaches. That exodus reduces both medical staffing and oversight of those who may be affected by the heat. Undoubtedly, the same weather conditions in July would have produced substantially fewer deaths.

The "theory" that allows climate change to be blamed for an increase in heat waves is that with global warming, climate will be more variable. Though the jury is still out worldwide on that, there is plenty of evidence for the United States that the opposite is true. A series of studies by Indiana University's Scott Robeson found that, for a large set of U.S. cities, places that have warmed most exhibited less temperature variability, not more. Regrettably, these and other key papers were not part of Patz et al.'s review.

Patz et al. also blame mortality on urban heat islands—the heat-trapping effects of buildings and paved surfaces combined with less vegetation—that result in most large cities' being significantly warmer than the surrounding countryside. They are correct. In fact, the urbanization effect can exceed the background rate of regional warming significantly, by one order of magnitude or more. So, if this is such a problem, we should expect people living in cities to be dying from heat exposure in increasing numbers.

Figure 6.3 shows the aggregate heat-related death rate toll for 28 of the largest U.S. cities from 1964 to 1998. There is a statistically significant decline in heat-related mortality over the period. During the same time, effective temperature (a combination of temperature and humidity) increased by an average of almost 1°C (1.8°F), mostly because of heat island effects. Why aren't more people, instead of fewer people, dying from heat exposure, as Patz et al. postulate?

It's simple. People, by and large, are not stupid. If it's too hot, they find an air-conditioned spot. If it's too cold, they turn up the heat, go out in the sun, or put on a jacket. The fact that Phoenix has a thriving population in a valley that is essentially inhospitable to pretechnological human habitation speaks volumes about human adaptability. Most elderly people move to Phoenix or Miami thinking in part they might prolong their lives (or at least live more comfortably) by living away from harsh winter weather, not so they could die sooner.

The review later waxes poetically about the potential health impacts of El Niño across the globe. Other horrors follow: epidemics

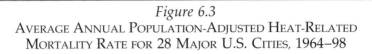

Figure 6.3
AVERAGE ANNUAL POPULATION-ADJUSTED HEAT-RELATED MORTALITY RATE FOR 28 MAJOR U.S. CITIES, 1964–98

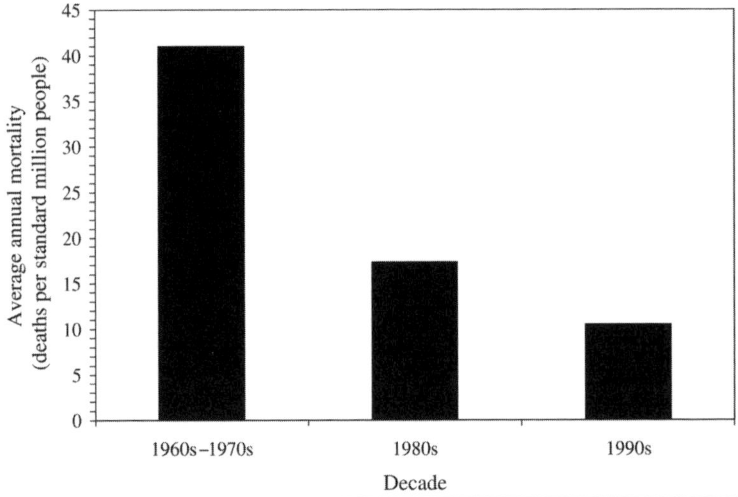

SOURCE: Adapted from Davis et al. 2003.

of malaria and Rift Valley fever, dengue hemorrhagic fever in Thailand, hantavirus pulmonary syndrome in the Desert Southwest in the United States, waterborne diseases in Peru, and cholera in Bangladesh.

One teensy problem . . . there's no clear causal link between El Niño (often acronymed ENSO for El Niño–Southern Oscillation, its true scientific name) and warming. If there were, it would be obvious by now. Patz et al. admit this (sort of): "Although it is not clear whether and how ENSO dynamics will change in a warmer world, regions that are currently strongly affected by ENSO . . . could experience heightened risks if ENSO variability, or the strength of events intensifies." Sure. An equally likely scenario is that the impact of all of these diseases will be reduced *if* global warming generates fewer and weaker El Niños.

Finally, the review paper pulls out a 3-year-old World Health Organization study and suggests that climate changes that have occurred in the last 30 years could have caused 150,000 deaths per year worldwide. On the basis of back-of-the-envelope calculations using current global population and mortality rate estimates, we determine that "global warming" is responsible for 0.2 percent of all deaths. This is a remarkably small number based upon WHO figures that are controversial in the first place.

Another way to look at global warming and mortality is on the benefits side. Since 1900, primarily as a result of technologies developed in a world powered by fossil fuels, average human life expectancy has increased significantly, doubling in the industrialized world. Doubling life expectancy is equivalent to saving one life. If two billion people lived during this period, then that would be equivalent to saving a billion lives. The World Health Organization's numbers don't bother to take that into consideration.

Noting that human-induced warming appears to "begin" around 1975, and that surely there were less than 150,000 excess deaths at the beginning, let's average the number of deaths per year from 1975 to 2000 at 75,000, giving a body count of roughly 1.9 million. Assuming that this 25-year period also includes 25 percent of the "saved" lives means 250 million "excess" lives. The net result after allowing for both deaths and "saves" is 248.2 million more living people.

The most important and interesting aspect of the *Nature* review article is that Patz, whose primary expertise is in vector-borne diseases such as malaria, and colleagues have the least confidence about the global warming–malaria link. Their discussions and review of the vector-borne disease literature are fairly balanced and contain many of the key caveats. Unfortunately, that balanced tone does not permeate most other aspects of the review.

There is no doubt that climate change will have some impacts, both positive and negative, on human health. One could just as easily write a review about how a warming planet is producing myriad health benefits.

Fewer French Fried in 2006

In the history of global-warming scare stories, the 2003 European heat wave was a lulu. But did you hear about the huge French heat wave of 2006?

You guessed it: many fewer people died. The 2006 heat wave is the subject of a recent paper in the *International Journal of Epidemiology* by a group of French researchers, led by A. Fouillet of the University

Figure 6.4
OBSERVED DAILY MORTALITY RATE AND MAXIMUM AND MINIMUM TEMPERATURES IN FRANCE, JUNE 1–SEPTEMBER 30, 2006

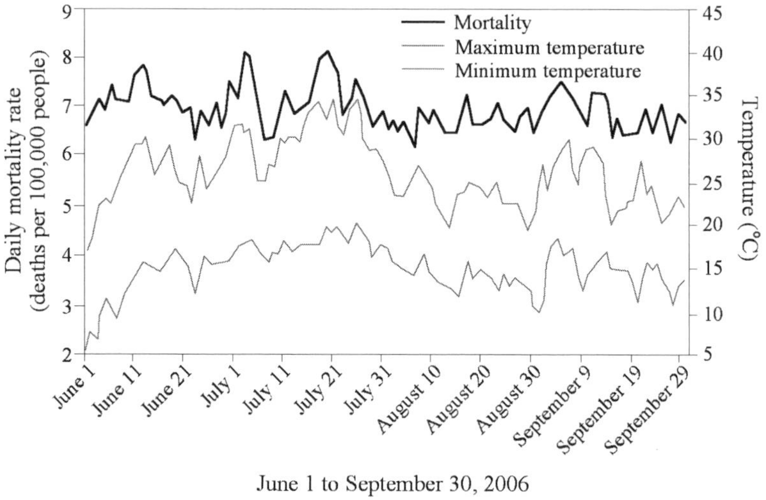

SOURCE: Adapted from Fouillet et al. 2008.

of Paris, titled, "Has the Impact of Heat Waves on Mortality Changed in France since the European Heat Wave of Summer 2003? A Study of the 2006 Heat Wave."

Fouillet and colleagues began by developing a model in which they used temperature data to predict daily summer mortality rates over the historical record (1975–99). There is a surprisingly strong relationship between temperatures and summer death rates (their overall model-explained variance was 79 percent). Figure 6.4 demonstrates this by comparing minimum and maximum temperatures and mortality rates for June through September 2006. It's easy to visually track the linkage between the temperature and mortality lines. Note during mid- to late July in particular, when the heat wave was going full blast, that mortality rates persisted at well over 7 daily deaths per 100,000 population.

Because of these very consistent linkages between temperature and mortality, the authors were able to statistically estimate the

Figure 6.5
Observed and Predicted Daily Mortality Rates in France, June 1 through September 30, 2003 (top), and June 1 through September 30, 2006 (bottom)

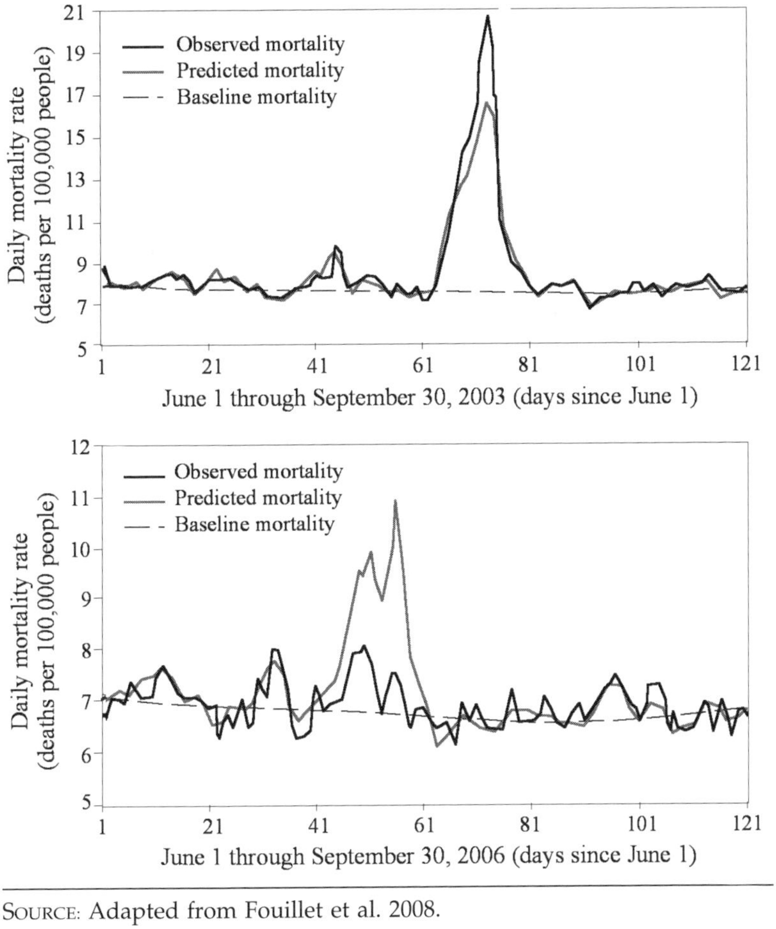

Source: Adapted from Fouillet et al. 2008.

number of deaths expected based upon observed temperatures. Figure 6.5 (top) shows the predicted and observed values for the notorious summer of 2003. It's not hard to find the heat wave in this graph. At the mortality peak, the model predicted about 17 deaths per 100,000, but nearly 21 deaths per 100,000 were observed. In other words, far more people died than would have been expected on the basis of the observed temperatures.

In July 2006, the situation was quite different (Figure 6.5, bottom). Now, the temperature model predicted far more deaths than actually occurred. Although there was a mortality spike, the model estimated almost 6,500 excess deaths during the heat wave, but only around 2,000 occurred. Specifically, the death count was 4,388 less than expected.

Can we attribute these saved lives to global warming? Well, maybe indirectly. In response to the tragedy of 2003, the French government implemented a National Heat Wave Plan that included surveillance of health data, recommendations on the prevention and treatment of heat-related morbidity, air-conditioning for hospitals and rest homes, and emergency plans for retirement homes, among other things. In other words, France *adapted* after the 2003 heat wave by providing information to the population at large and air-conditioning to the most vulnerable. No doubt people were also personally more aware of the dangers of summer heat in 2006 than they were three years earlier. In reality, there is no excuse for such mass heat-related mortality that occurred in 2003 in any technologically advanced country.

Let's face it. The planet is warming, and, short of China and India going green, there's nothing anybody can really do about it with current technology. (Go ahead and install energy-saving light bulbs, but it's not going to do much to change global temperatures.) It would be foolish to argue that we're not going to see more and longer summer heat waves as greenhouse gas levels continue to increase. But that certainly doesn't mean that death rates are going to skyrocket. In fact, with some relatively simple adaptive measures, death rates could very well go down regardless of future temperatures. That's been the trend in the United States and should be the standard throughout the developed world.

Science Fiction: The Imminent Ice Age

Atlantic's Salt Balance Poses Threat, Study Says

> The delicate salt balance of the Atlantic Ocean has altered so dramatically in the last four decades through global warming that it is changing the very heat-conduction mechanism of the ocean and stands to turn Northern Europe into a frigid zone.

> The conclusions are from a study in the journal *Nature* that is to be published today. The study describes planet-scale changes in the regulatory function of the ocean that affect precipitation, evaporation, fresh-water cycles, and climate.
>
> This has the potential to change the circulation of the ocean significantly in our lifetime," said Ruth Curry of the Woods Hole Oceanographic Institution in Massachusetts, the study's lead author.
>
> —*Toronto Globe and Mail*, December 18, 2003

Who hasn't read something similar to this quotation from the Toronto paper? Or seen headlines such as "Global Warming to Cause Next Ice Age!" or "Global Warming to Send Europe into a Deep Freeze!" In fact, next time New England or Europe has a cold winter, it's a guarantee that you'll see them again. Why? Because history is somewhat repetitious.

Nor are these stories confined to a normally left-leaning press. For example, New England saw a pretty decent and prolonged cold snap in January 2003, prompting *Wall Street Journal* science writer Sharon Begley to proclaim:

> The juxtaposition of a big chill in the Northeast and near-record warmth globally seems eerily like the most dire predictions of climate change: As most of the world gets toastier, average winter temperature in Northeastern America and Western Europe could plunge 9°F.

The idea behind these scary stories (and the premise of the monumentally bad science fiction film *The Day After Tomorrow*) is that the ocean's "thermohaline" circulation slows down or, even worse, stops, sending the climate into disarray—all because of anthropogenic global warming. In the case of *The Day After Tomorrow*, this circulation shutdown led to a flash freeze of the planet. Peter Schwartz and Douglas Randall, two nonclimatologist contractors to the U.S. Department of Defense said, in an October 2003 report, that this could happen in the next decade! Their highly amusing (to climatologists) report was titled, "An Abrupt Climate Change Scenario and Its Implications for United States National Security."

The thermohaline (thermo = heat; haline = salt) circulation works like this: Strong solar heating and warm waters in the tropical Atlantic result in enhanced evaporation there, leaving the surface waters there saltier than the average ocean. These warm, salty waters are carried northward via the Gulf Stream, and in the high latitudes they release their heat into the atmosphere, and subsequently cool. This cool, salty current of water becomes more dense than the less salty waters surrounding it and consequently sinks and flows back southward, acting as a sort of pump that drives this major circulation system that circuitously winds its way though most of the world's major oceans.

There are indications from paleoclimatological studies that the thermohaline circulation has shut down in the past, causing "abrupt" change. This happened 8,200 years ago, when the world was still emerging from the last ice age. As the ice sheet that covered North America melted, it formed a huge lake in central Canada (Lake Agassiz) that contained more water than the combined Great Lakes do currently. Lake Agassiz was held back by an ice dam that eventually disintegrated as the climate warmed, and immediately the lake's contents roared through Hudson Bay and into the North Atlantic Ocean. This fresh "meltwater pulse" apparently shut down the pump, which took a couple hundred years to get up and running again, during which time Greenland and Europe were considerably cooler, around 5.5°F (3.1°C), than they were prior to this event.

Today, there is no bigger-than-all-the-Great-Lakes-combined glacial meltwater lake in central Canada held back by an ice dam on the verge of collapse. But that hasn't led folks to abandon the idea that human-induced climate warming may cause a shutdown of the ocean's thermohaline circulation. The idea is that global warming will lead to a meltdown of the Northern Hemisphere's last significant ice sheet remnant from the Ice Age—the one lying atop Greenland. The meltwater from the Greenland ice sheet, together with a projected enhancement of high-latitude precipitation, will eventually provide a large enough input of fresh water to the subpolar North Atlantic ocean to slow and eventually halt the thermohaline circulation.

As a result, researchers have been poring over data in search of any indications that that is happening. Any hint that they may have identified a thermohaline slowdown produces a rasher of lurid stories. For example, in December 2003, Ruth Curry, from Woods Hole Oceanographic Institution, and colleagues reported in *Nature* magazine that they had detected evidence that suggested the thermohaline circulation was slowing down. From data collected from the 1950s to the 1990s, they reported that the tropical Atlantic was growing saltier while the northern latitudes of the Atlantic were growing fresher and suggested that anthropogenic global warming was the probable cause. That prompted the *Toronto Globe and Mail* article.

Then, two years later, in December 2005, Harry Bryden and colleagues published an article, again in *Nature* that seemed to show additional evidence of the thermohaline circulation's long, slow death march. They examined a series of ship transects of the Atlantic Ocean, the first in 1957 and the last in 2004, and declared that they had detected a circulation slowdown of about 30 percent, which primarily had taken place during the past 10 to 15 years. That prompted this headline and story:

> Alarm Over Dramatic Weakening of Gulf Stream
>
> The powerful ocean current that bathes Britain and northern Europe in warm waters from the tropics has weakened dramatically in recent years, a consequence of global warming that could trigger more severe winters and cooler summers across the region, scientists warn today.
>
> Researchers on a scientific expedition in the Atlantic Ocean measured the strength of the current between Africa and the east coast of America and found that the circulation has slowed by 30 percent since a previous expedition 12 years ago.
>
> —*The Guardian*, December 1, 2005

In the intervening time, more and better data have come in, prompting articles demonstrating that it is impossible to verify a slowdown in the thermohaline circulation related to anthropogenic influences. Instead, they conclude that natural variations in the strength and speed of the thermohaline circulation can explain the observed behavior.

The latest of these appeared in 2007 in *Geophysical Research Letters*, by Tim Boyer and colleagues, who hail primarily from the U.S. National Oceanic and Atmospheric Administration (NOAA). Boyer et al. examined salinity trends in the waters at various depths and latitudes covering the North Atlantic Ocean. They found, overall, a general salinity increase over the entire basin. As shown in Figure 6.6, although the waters of the tropical latitudes showed a fairly steady trend toward enhanced salinity between the beginning of their study period (1955) and the end (2006), northern waters showed a clear freshening trend lasting from about the late 1960s

Figure 6.6
EQUIVALENT FRESHWATER CONTENT FOR DIFFERENT AREAS OF THE NORTHERN ATLANTIC OCEAN, 0–2,000 METERS, 1955–2006

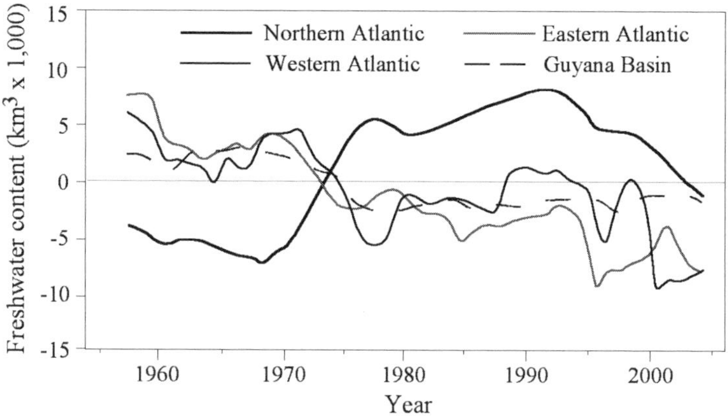

SOURCE: Adapted from Boyer et al. 2007.

NOTE: 2,000 meters = 1.24 miles; km^3 = cubic kilometer.

through the early to mid-1990s. Since then, the freshwater content of the northern latitudes of the Atlantic Ocean has been *declining*.

The freshening trend of the high-latitude Atlantic coupled with increasing salinity in the lower latitudes undoubtedly made it seem, at least when it was occurring, that the thermohaline circulation was slowing down as a result of global warming—as reported by Curry et al. in 2003. But their data ended in 1999, which was too early to pick up the end of the freshening trend and the increasing salinity. All of that indicates that the thermohaline circulation is still quite healthy, and in fact, that it has likely strengthened over the past 10 years or so—counter to the suggestions of Curry et al. and Bryden et al. in 2005.

Interestingly, there may be a tie-in to Atlantic hurricane activity. For years, hurricane guru William Gray has been saying that the strength of the thermohaline circulation is an important determinant of Atlantic hurricane activity.

Boyer et al. seem to give added credence to Gray's idea. If you take the freshwater content of the northern portions of the North Atlantic as an indicator of thermohaline strength (fresher is weaker, saltier is stronger), there's a strong association between circulation strength (Figure 6.6) and the number and strength of Atlantic hurricanes (Figure 3.3).

From the late 1950s through the early 1970s, when the high-latitude, subpolar Atlantic waters were relatively salty, there were plenty of hurricanes and a lot of strong ones. As the subpolar Atlantic freshened from the early 1970s to the mid-1990s, Atlantic hurricane activity diminished. Hurricane activity once again picked up in intensity and frequency in 1995 at precisely the time when the subpolar waters began to increase in salinity. Because salinity changes evolve over several years, the current hurricane regime is likely to continue.

So, we either get a strong hurricane season (which will be blamed on global warming) with a healthy thermohaline circulation, or weak and infrequent hurricanes, with a sick thermohaline circulation (which will be blamed on global warming). What a win-win situation for those fanning global warming fears.

Extreme Heat and Cold in the United States

The previous chapter detailed Ken Kunkel's work on the frequency of extreme rainfall in the U.S., where he found that the recent era sure looks a lot like it did 100 (colder) years ago. Kunkel exploited newly digitized Cooperative Observer data from 1895 through 1948.

Using this expanded set of climate data, Kunkel also examined the frequency and magnitude of heat and cold waves, defined by a 4-day event with an average appearance of once every five years. The heat wave record (Figure 6.7, bottom) is dominated by the huge spike during the 1930s "Dust Bowl" era. Recent heat is hardly noticeable in the longer-term context, even though the number of heat waves has increased recently after the cool summers of the 1960s and 1970s.

For cold waves (Figure 6.7, top), you'd think that would be a no-brainer in greenhouse world. After all, we showed in chapter 1 that the coldest nights of the year are warming more than any others. Most climatologists (including us) believe they are arguing sensibly when they say we should expect fewer cold waves as our winter air masses warm.

Data induce humility. Figure 6.7 shows no obvious signal in the frequency of cold waves.What about arguments that global warming

Figure 6.7
U.S. COLD WAVE (TOP) AND HEAT WAVE (BOTTOM) FREQUENCY INDICES, 1895–1997

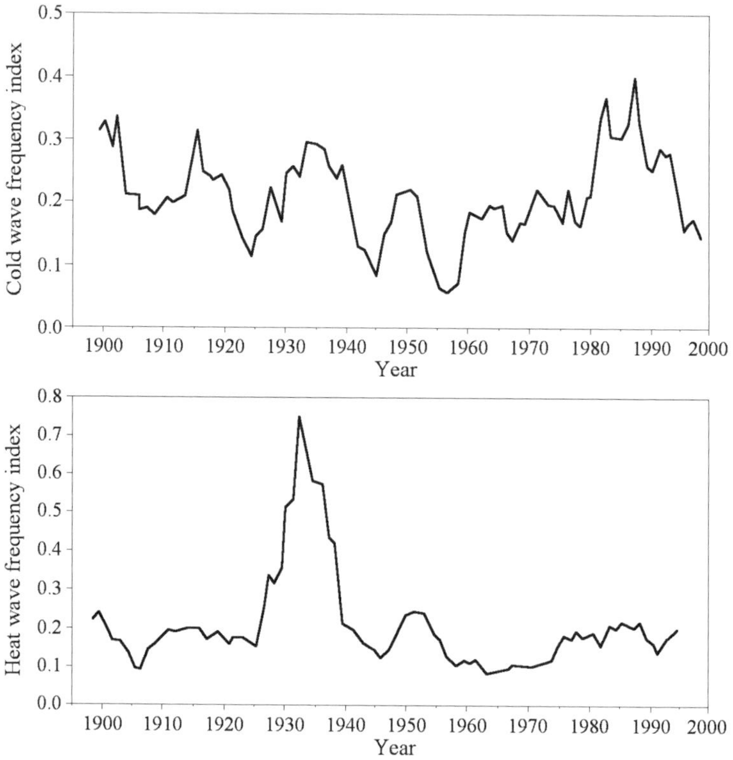

SOURCE: Kunkel 2006.

will produce more extreme heat *and* cold? Kunkel's data support none of this. If more cold waves are the result of global warming, why have peaks that dominated the 1980s completely disappeared? And if we should expect fewer cold outbreaks, then how does one account for all the cold air outbreaks in the 1980s when the atmosphere had plenty of greenhouse gases? The cold wave record clearly shows some interesting long-term variability but no obvious trend.

Those data, from the lower 48 states, are for less than 2 percent of the earth's surface. But the United States has the most dense long-term weather records of any similar-sized place on the planet. Note how the addition of the pre-1948 Cooperative Observer figures can completely change one's interpretation of trends (which we also showed for extreme precipitation in the previous chapter).

As with all issues of global climate change, the devil's in the details, but details about weather records hardly make for blockbuster headlines.

7. Pervasive Bias and Climate Extremism

The story portrayed in this book is that there is a body of science—an internally consistent one—that paints a picture that is much different than the gloom-and-doom vision of climate change that we read about almost every day. The material should prompt several questions, including the obvious one: "Why haven't I heard about this?"

This chapter is an attempt to explain. It does not provide a definitive answer, because we believe that the answer is complex and currently unknown in any universal sense. But there are known aspects of the behavior of science itself, and of the public presentation of scientific information, that certainly have some explanatory power.

The larger question is this: How did we get to global warming's climate of extremes? Is it a product of the reporting of science, or is it because of the behavior of scientists themselves and global warming science itself?

Certainly, few scientists would blame themselves. But, it turns out, what climate scientists believe to be true about their own global-warming research is not in fact true. Climate scientists believe that there are no inherent biases in their work as it is published in the scientific literature.

If bias is inherent in the science itself, then that primary information stream that is fed to the media is itself biased. Even if the journalistic community were philosophically, rhetorically, and editorially neutral, the bias would come shining through.

That is best demonstrated with an example from something with very little political baggage: the daily weather forecast.

Science as an Unbiased Penny

Step away from forecasts of the temperature 100 years from now. Go to more familiar territory: the weather forecast several days ahead.

As of this writing, on a pleasant Sunday in May, the forecast high temperature for next Thursday here in Washington, DC, is 84°F (29°C). That forecast was generated by a computer model, based upon a worldwide, simultaneous, horizontal and vertical snapshot of the physical atmosphere taken at 8:00 a.m. Eastern Daylight Time (EDT) today.

Every day, two of these worldwide snaps are taken simultaneously: at noon Greenwich Mean Time (GMT) (named for Britain's Greenwich Observatory) and at midnight GMT (8:00 a.m. and 8:00 p.m. EDT, respectively).

Four hours from now, today's second observation will be taken, and it will take an additional hour and a half to be input to and to run through the world's various weather forecasting models.

Although some of the models will predict the original 84°F, others will predict some other temperature. After discounting those models that didn't change, what are the probabilities that the new forecast will be either warmer or colder than the original 84°F?

Exactly equal. If we assume there wasn't something systematically wrong with the original forecast—meaning that bad data or calculation errors made it too warm or too cold—the probability of a new forecast raising or lowering the result from the previous one is the same.

Weather forecasters believe that, and so do climate forecasters. They literally took this claim all the way to the U.S. Supreme Court (more on that in a moment). And they were wrong. The world of climate science turns out to be naturally biased.

Yet the belief in nonbias permeates the community. That's what David Battisti and 18 other people, calling themselves "the Climate Scientists," showed in an amicus brief concerning the first case relating to the regulation of greenhouse gases (*Commonwealth of Massachusetts v. U.S. Environmental Protection Agency*). Indeed, "the Climate Scientists" included some very big names, such as NASA's James E. Hansen, head of the Goddard Institute for Space Studies, and Nobel Prize winners Mario Molina and F. Sherwood Rowland.[1]

[1] In an amicus curiae (Latin for "friend of the court"), authors are always listed alphabetically, rather than by rank or by the amount of contribution to the brief; amicus briefs are submitted to offer information intended to inform the court's decisionmaking.

Equal Justice Under Law for Climate Scientists?

"Equal Justice Under Law." Those words are carved into the frieze of the U.S. Supreme Court building in Washington—but do they apply to climate scientists' freedom of speech?

NASA's James Hansen was a coauthor of Alan Battisti et al.'s brief in *Massachusetts v. U.S. Environmental Protection Agency*, which maintained that the preponderance of scientific evidence demonstrated that global warming caused by carbon dioxide had such potential for environmental damage that it fell under the 1990 Amendments to the Clean Air Act, requiring the EPA to issue rules and regulations. In the section of the brief labeled, "Interests of the *Amici Curiae*," Hansen describes himself as director of NASA's Goddard Institute for Space Studies. Therefore, in expressing that claim, Hansen was siding against his employer, the federal government, which administers both NASA and the EPA.

Harvard's Sallie Baliunas et al. wrote another amicus brief countering "the Climate Scientists," in which they argued that there was no extant comprehensive study of the full net effect of carbon dioxide–induced global warming, and that therefore there was no technical basis for regulation. She and her fellow amici, including Delaware State Climatologist David Legates and the authors of this book, were therefore siding with the federal government.

Delaware was well represented, albeit on both sides. Delaware Attorney General Carl Danberg had entered a separate brief in support of Massachusetts (i.e., against the federal government). As a result, two entities that would both appear to represent the state of Delaware had come down on opposite sides of the case. As this all came to a head, Delaware Governor Ruth Ann Minner (D) told Legates that he could not speak about global warming using the title of State Climatologist. Hansen, however, received no notification from NASA that he could not refer to himself as director of the Goddard Institute (an obviously federal entity) when giving his views on global warming, even though he argued *against* the federal case.

In discussing the effects of carbon dioxide–induced warming on human health, the Battisti et al. brief said:

> EPA also ignored the two-sidedness of scientific uncertainty. Outcomes may turn out better than our best current predictions, but it is just as possible that environmental and health damages will be more severe than the best prediction.

The key words are "just as possible." What Battisti et al. are claiming is that each new research finding stands an equal probability of making the predicted effects of global warming on "environmental and health damages" "better" or "more severe."

Battisti et al. are contending that updated climate science is analogous to updated weather forecasts—that is, that there is an equal probability that new results would be either more severe or more moderate.

Battisti et al. would certainly claim to represent the mainstream of scientific opinion on this issue. The primary citation in their amicus brief is a 2001 report on climate change from the U.S. National Research Council titled *Climate Change Science: An Analysis of Some Key Questions*. What could be more mainstream then the National Research Council?

Battisti et al. actually framed a very testable hypothesis, by assuming that new findings have an equal chance of projecting either the warming or its effects "better" or "worse."

If their hypothesis is true, this climate research community must believe that there is no "publication bias" in climate change research in either a positive or a negative direction. Such an assumption is natural for atmospheric scientists, given that many of them are trained in weather forecasting. In general, forecast models are corrected for internal biases to the fullest extent possible before they become operational. In fact, the quantitative output of weather models, called Model Output Statistics, are mathematically constrained to be unbiased in any direction. So each new iteration of a forecast model does indeed have an equal probability of moving the output statistics up or down.

If climate models are similarly unbiased, then new information on forecast climate change or its impact should also have an equal probability of making existing forecasts worse or better. On the other hand, the assumption will be untrue if the consensus of extant forecasts is simply wrong with regard to the severity or lack of warming because of some basic flaw in model concept or design.

Still, there are a lot of climate models out there, all supposedly of independent design. There have now been four studies "intercomparing" them under common conditions (i.e., similar changes in carbon dioxide), and (the first is shown as Figure 7.1; see insert) they tend to behave similarly. Consequently, there is no compelling reason to believe in some pre-existing, pervasive model bias that needs correction in one direction.

So, at first blush, the "consensus" that there is no bias would seem to be reasonable. But economists and biomedical scientists would disagree. They have written an extensive literature on "publication bias."

How does the scientific literature become biased in one direction? One cause has nothing to do with any personal bias or self-interest on the part of the scientists. Rather, it has to do with the nature of scientific "news."

Negative results are generally considered not noteworthy. For example, if a researcher "discovered" that there was no relationship between, say, regional birth rates and global warming, no scientific journal is likely to accept such a paper because (presumably) no relationship was expected.

In other words, scientific journals are skewed by a prejudice for the publication of statistically significant, "positive" results and prejudiced against findings of no relationship between hypothesized variables.

Harvard's Robert Rosenthal is generally thought to have published the seminal paper in 1979 on this type of publication bias, which he called the "file drawer problem." According to Rosenthal:

> For any given research area, one cannot tell how many studies have been conducted but never reported. The extreme view of the "file drawer problem" is that journals are filled with the 5 percent of the studies that show Type I errors [a "positive" result], while the file drawers are filled with the 95 percent of the studies that show nonsignificant results.

Rosenthal even went on to develop a quantitative methodology to correct for "missing" negative results when analyzing large numbers of papers dealing with a specific topic. The notion is quite important in the biomedical literature, where there can be quite clear

Peer Review, Bias, and the Refereed Literature

The canon of science consists of the peer-reviewed or "refereed" literature. In the classic sense, the process is rather straightforward.

A scientist submits an article to a peer-reviewed journal, such as *Science*, *Nature*, or *Geophysical Research Letters*. The editor first decides if the material is appropriate in general, and then sends it to two or three outside reviewers for comment. Those reviewers—who are peers from the author's professional community—advise the editor whether to publish the manuscript in its original form, to accept it with modification, to reject it but to entertain another submission with modification, or to reject it outright.

Ideally, the name(s) of the authors should not be made available to the reviewers, but that practice has long since vanished, at least in climate science. Further, the reviewers should have no particular interest whether or not the submitted manuscript sees publication, and they should have a considerable degree of professional independence from the writers of the manuscript in question.

In a 2007 book, *Controversy in Global Warming: A Case Study in Statistics*, statistician Edward Wegman expressed concern about the possibility of bias within the peer review process: "In particular, if there is a tight relationship among the authors, and there are not a large number of individuals engaged in a particular topic area, then one may suspect that the peer review process does not fully vet papers before they are published."

Wegman was interested in the scientific review of papers by Pennsylvania State University's Michael Mann, who published several articles purporting to show that the warmth of the second half of the 20th century was unprecedented in at least the last millennium. The principal result is a graph widely known as the "hockey stick" plot because it shows very little change for 900 years (the handle) and then a sudden jump in the last century (the blade).

(continued on next page)

(continued)

Wegman found that it was highly likely that the reviewers of Mann's work were also people with whom he had coauthored other papers, and so the probability of an independent review was greatly diminished.

And what's really left of peer review, anyway? The American Geophysical Union asks people who submit scientific papers to provide the names of five people they think would be desirable reviewers and also (optional) a list of people that the author belives could not provide an objective review. When the writer can influence the selection of reviewers, peer review is pretty much dead.

In other words, authors can now tell editors who their friends are and who can't be "objective." This sounds like a formula for publication bias.

consequences for nonpublication of negative results. People may continue drugs or treatments that in fact have no effect on disease progression. But there seems to be no discussion or any documentation of publication bias in the climate change literature.

Rosenthal's model should apply to climate change. Perhaps there is a bias toward "positive" manuscripts' relating global warming to some effect, and a bias against results where there is no relation.

A related logic applies to the magnitude or importance of some phenomenon. In global warming, are there an improbable number of publications indicating that warming, or its effects, will be greater (i.e., "worse than expected") rather than more modest (i.e., "not as bad as expected")?

The famous polymath Stephen Jay Gould described another cause of publication bias in 2002. He stated that publication bias results from "prejudices arising from hope, cultural expectation, or the definitions of a particular theory [that] dictate that only certain kinds of data will be viewed as worthy of publication, or even documentation at all." Gould's definition is quite analogous to the late Thomas Kuhn's notion of a scientific paradigm, first elucidated in his 1962 classic, *The Structure of Scientific Revolutions.* In Kuhn's view, paradigms are overarching logical structures that provide the best explanation of a family of phenomena. When interfaced with publication bias, a paradigm should become increasingly defensive and exclusionary.

The body of scientific work that makes up the peer-reviewed literature is what defines a reigning paradigm. Are there other processes that intrude in this literature that go beyond the "file drawer" and paradigm effects?

Consider the dynamics of "Public Choice Theory," first described in 1962 by Nobel Prize winner James Buchanan and Gordon Tullock. They argued that "individuals will, on the average, choose 'more' rather than 'less' when confronted with the opportunity for choice in a political process, with 'more' and 'less' being defined in terms of measurable economic position."

In the case of global warming, public choice influence may occur at several levels that could promote publication bias. As virtually all global warming science is a publicly funded enterprise, political dynamics must in part be involved. At the simplest level, global warming is just one of many scientific issues competing for funding. AIDS and cancer, for example, are competitors.

Because the scientific budget is finite, the perceived importance of each competing issue determines in part how much support each one receives. It is difficult to imagine, at the level of Congressional hearings, that high-level managers or funding recipients for any of these issues would dare portray them as relatively unimportant. That creates a culture in which any scientific finding undermining importance (or in this case, indicating less, or less costly, climate change) becomes economically threatening. As a result, the peer review process would *have* to become biased, unless reviewers never act in their own interest.

Further, the reward structure in academia—promotion, tenure, and salary—is based on the quality and quantity of peer-reviewed research. The requisite level and number of publications for tenure is virtually impossible to achieve without substantial public funding. Interestingly, Julia Koricheva, of the Royal Holloway University in London, found in 2003 no evidence that dissertation-related publications suffered from publication bias. Perhaps, then, it is a correlate of professional development.

This funding stream, and therefore career advancement, is threatened by findings downplaying the significance or magnitude of climate change. Under this model, articles submitted for publication making the case for less-than-alarming findings would likely receive more vigorous and negative reviews than articles arguing otherwise.

Stanley Trimble, a geographer at UCLA, recently summarized what he calls "the double standard in environmental science." He wrote (in the Cato Institute journal *Regulation*) that, in the major journals, specifically and repeatedly citing *Science* and *Nature*, "The implication is that flimsy or even no evidence for environmental degradation is acceptable, but any evidence for improvement is suspect."

Further:

> For those of us in academe, all of this can have profound implications for careers that depend upon having many (but not necessarily *good*) publications for advancement. And based upon prima facie evidence, environmental extremism is good for the career.

In summary, there are at least three social processes that argue against the assertion of "the Climate Scientists" that new research results have an equal probability of describing a more or less severe effect from global warming: The file drawer problem, the paradigmatic nature of science, and public choice dynamics.

While there are no obvious citations concerning the existence of publication bias in the atmospheric science, meteorological, or climatological literature, it has been discussed in evolutionary ecology (i.e., Gould's 2002 statement), and by University of Birmingham's (U.K) Phillip Cassey et al., in 2004 in *Proceedings of the Royal Society*. Cassey et al. noted that analyses of collections of papers ("meta analyses") assume that the original work is nonbiased. In their study, Cassey et al. found pervasive bias against negative results related to sex ratios.

Cassey et al. concluded that "publication bias is not just a 'file drawer' problem but can also be manifest within the primary literature. This is particularly likely to be a problem in research fields where results are presented for a large number of independent test variables, such as in ecology." It certainly seems reasonable that climate change research (and its interface with biology) would behave analogously. Clearly, the nexus of interaction between climate and biology is ecology.

Determining whether the public choice dynamics, the cultural expectation–derived bias, or the file drawer problem, or some combination of the three is creating publication bias in climate science is a daunting task.

But, before undertaking that, we need to establish whether or not publication bias exists in the world of climate change.

That's an easy determination. Assume that each new piece of information that contributes to some forecast of a future phenomenon (or the effects of one) should have an equal probability of making that forecast "worse" or "better." Documentation of a highly improbable distribution of "worse" and "better" would support the hypothesis that publication bias in climate change research is a real phenomenon.

Let's do a little experiment, examining climate change–related articles in the journals *Science* and *Nature* from July 1, 2005, through July 31, 2006. In *Science*, let's look at the sections titled "Perspectives," "Reviews," "Brevia," "Research Articles," and "Reports." In *Nature*, we'll do "News," "News Features," "Correspondence," "News & Views," "Articles," and "Letters." A total of 116 relevant articles were counted in this period, or slightly more than two per week (or one per magazine per week)—52 in *Science* and 64 in *Nature.*

Let's place each article into one of three classes, labeled for convenience as "better," "worse," or "neutral or could not classify."

Articles were placed in the "better" bin if the findings reduced the amount of prospective global or regional warming or unfavorable weather or climate, or reduced some impact or effect that had been previously established in the scientific literature. Similarly, articles in the "worse" bin increased the amount of global or regional warming, increased the frequency and/or magnitude of unfavorable weather or climate events, or increased the impact or effect of climate change on some responding phenomenon.

Articles were placed in the "neutral or could not classify" bin if they had approximately offsetting "better" and "worse" implications, or simply were not classifiable because of content.

Of the 116 articles, 84 were "worse," 10 were "better," and 22 were "neutral." The results are summarized in Table 7.1.

Some of the 116 articles were fairly easy and straightforward to classify. For example, King et al. (2006) wrote, "Acclimation of plants

Table 7.1
CLASSIFICATION OF ARTICLES ON CLIMATE CHANGE IN *SCIENCE* AND *NATURE*, JULY 2005–JULY 2006

Journal	"Better"	"Worse"	Neutral	Total
Science	5	34	13	52
Nature	5	50	9	64
Total	10	84	22	116

SOURCE: 116 articles in *Science* and *Nature*, July 1, 2005, through July 31, 2006; a full bibliography with classifications appears at the end of this book.

to higher temperatures may reduce the extra warming caused by increased plant respiration in a future warmer world." That clearly belongs in the "better" bin. Similarly, when Schimel (2006) titled an article "Climate Change and Crop Yields: Beyond Cassandra," and wrote that "previous studies overestimated the positive effects of higher carbon dioxide concentrations on crop yields," there's little debating that it goes in the "worse" bin.

But some weren't very straightforward at all, indicating both "worse" and "better" aspects within the same publication. Classification was made on the basis of whether one or the other dominated. If neither did, the report was classified as "neutral." In some cases, though the aspects were mixed, one side outweighed the other, as with Feddema et al. (2005) on land-use influences on climate change; their article included both positive and negative effects, but there were more negative than positive, so the paper was classified as "worse." An example of a "neutral" result from competing effects can be seen in Gedney et al. (2006), who, writing about water budgets and carbon dioxide, concluded, "As the direct CO_2 effect reduces surface energy loss due to evaporation, it is likely to add to surface warming as well as increasing freshwater availability." Sounds like some good, some bad.

"Neutral" reports included those that were consistent with previously published either negative or positive effects. Consequently, a paper by Field et al. (2006) finding biological evidence for deep-ocean warming is "neutral," given the previous large body of work (Barnett, Pierce, and Schnur 2001; Levitus et al. 2000; etc.) showing the same results. Note that such labeling does not mean that the original background work was necessarily neutral, only that it was published before the beginning of our examination period.

Timing and Publication Bias

Two concurrent events in December 2005 exemplified how publication bias and news go hand-in-glove. They were the Montreal "Conference of the Parties" that had signed the UN Kyoto Protocol on global warming and the fall meeting of the American Geophysical Union (AGU) in San Francisco.

The sheer volume of hype was impressive. Following are the headlines, along with the sources, generated on the afternoon of December 7, first from the Montreal UN conference. (University news sources are those that were eventually picked up in other stories). These were obtained from Google's news search page.

- "Global Warming to Halt Ocean Circulation" (University of Illinois)
- "Warming Trend Adds to Hazard of Arctic Trek" (*Salem [Oregon] News*)
- "Pacific Islanders Move to Escape Global Warming" (Reuters)
- "Tuvalu: That Sinking Feeling" (PBS)
- "World Weather Disasters Spell Record Losses in 2005" (*Malaysia Star*)
- "Arctic Peoples Urge UN Aid to Protect Cultures" (Reuters)
- "Threatened by Warming, Arctic People File Suit Against US" (Agency France Press)

Next, from San Francisco:

- "Ozone Layer May Take a Decade Longer to Recover" (*New York Times*)
- "Earth is All Out of New Farmland" ([*London*] *Guardian*)
- "Forests Could Worsen Global Warming" (UPI)
- "Warming Could Free Far More Carbon from High Arctic Soil than Earlier Thought" (University of Washington)
- "Rain Will Take Greater Toll on Reindeer, Climate Change Model Shows" (University of Washington)

(continued on next page)

(continued)

- "Methane's Impacts on Climate Change May Be Twice Previous Estimates" (NASA)
- "Average Temperatures Climbing Faster than Thought in North America" (Oregon State University)

How can things be so bad?

Each one of those stories carries an "it's worse than we thought" subtext. There was a single additional story to the contrary, published by AP, that indicated that plants may store more carbon dioxide than was previously thought, which would help to limit warming.

Similarly, many observational studies, such as ice core research published in 2005 by Siegenthaler et al., are largely extensions of previous work. Unless such studies revealed behavior that was inconsistent with or largely different from related research, they were classified as "neutral."

The assumption of nonbias of "worse" vs. "better" is analogous to flipping a coin. The observed better/worse ratio has the same probability as flipping a coin 94 times and getting 10 or fewer heads (or tails). That would arise by chance in an unbiased sample with a probability of less than 5.2×10^{-16} (or a chance of less than 1 in 50,000,000,000,000,000).

Trimble writes that "*Science* and *Nature* are truly major gatekeepers of science; indeed they are the gold standard." Rather than performing a time-sequence analysis, as was done here, he examined their handling of soil erosion studies. He found a similar bias toward "worse" results, and concluded that

> [*Science* and *Nature*] have a special obligation to objectivity and even-handedness. But it seems clear that they sometimes have not maintained their charge as it pertains to environmental science.

An alternative explanation to publication bias is simply that the magnitude and impacts of warming have been systematically underestimated, and that the recent literature we examined is merely reflective of science that "self-corrects."

But the self-proclaimed climate community in *Massachusetts v. EPA* clearly believes that previous work is unbiased; otherwise, that group would hypothesize no equal probability that a new result might appear better or worse than previous ones. And a much broader modeling community, in the 2007 "Fourth Assessment Report" of the United Nations Intergovernmental Panel on Climate Change, makes a clear claim that modeled and observed climate changes over the last 50 years are very similar:

> It is *likely* that there has been significant anthropogenic warming over the past 50 years averaged over each continent except Antarctica. . . . The observed patterns of warming, including greater warming over land than over the ocean, and their changes over time, are only simulated by models that include anthropogenic forcing.

Consequently, both Battisti et al. in *Massachusetts v. EPA* and the larger IPCC community feel that the models are unbiased and that therefore new results have an equal probability of making the future appear "worse" or "better."

Obviously, the publications of the scientific community that are selected by *Science* and *Nature* have considerable influence and penetration into the media. So why is the news almost always bad? It's not necessarily because of journalistic bias. Instead, their primary source—the science *itself*—is biased in one direction.

The demonstrable existence of publication bias in global warming science has several consequences, especially when one tries to "summarize" climate science.

Repeated attempts have been made to use the primary literature to form such overall summaries or "state of the science" reports. The highly cited 2007 IPCC "assessment" of climate change is a primary example. Periodically, the U.S. National Research Council addresses climate change issues, and those reports largely rely on the published literature. (As noted earlier, a recent Council report served as much of the basis for Battisti et al.) There is never a discussion of the possibility of publication bias in these reports.

The consequences of synergy between publication bias, public perception, and scientific "consensus" and policy are very disturbing. If the results shown for *Science* and *Nature* are in fact a general character of the scientific literature concerning global warming, our climate-related policies are based upon a directionally biased stream of information, rather than one that has a roughly equiprobable distribution of altering existing forecasts or implications of climate change in a positive or negative direction. That bias exists despite the stated belief of the climate research community that it does not.

Note that bias does *not* mean the exclusion of differing points of view; it just means that, in the case of global warming, published research papers saying that warming will be attenuated or that the effects will be muted are likely to be fewer and further between.

To survive the minefield of publication bias, papers arguing for less global warming or less impact of climate change that appear in the refereed literature are likely to be compelling and relatively free of the major flaws. Although the ratio of alarmist to "worse" to "not so bad" is about 10 to 1, it is clear that there *is* a body of literature out there that may argue very cogently against the current hysteria.

The Internet and Peer Review

The Internet is dramatically changing the sacrosanct nature of the peer-reviewed literature. High-profile papers appearing in major journals hit the blogs, and are sometimes found to be deeply flawed after publication. Perhaps an example from something slightly less incendiary than climate change—stem cell research—will illustrate the process.

In March 2004, *Science* published an earthshaking paper by H. S. Hwang and several coauthors who claimed to have obtained stem cells from a cloned human embryo. It included several pictures.

Skeptical scientists detected something wrong with the images. They noted that the individual pictures actually were overlapping images of a single larger picture. Moreover, they had been seen before. Another team of researchers had submitted them to the journal *Molecules and Cells* before Hwang's paper was sent to *Science*. The earlier paper described them as cells that were created without cloning.

The *Boston Globe* had been following this issue on the blogosphere. It took the photos and sent them to several stem cell researchers, who pronounced them as identical.

Eventually, and with great embarrassment, *Science* retracted the paper. Donald Kennedy, the editor of *Science*, defended the review process, saying that reviewers could not be expected to detect deliberate falsehoods. Further, Kennedy noted that reviewers can demand more data if they are suspicious.

Rather than let the reviewers (and *Science*) try to slide off the hook on the stem cell fiasco, Kennedy should have thanked the Internet!

Steve McIntyre, himself a tireless and meticulous investigator of global temperature histories, cites the "Hwang Affair" as (yet another) example of poor peer review. In his blog *Climate Audit* (http://www.climateaudit.org), he noted that

> While Western scientists were still supporting Hwang, young Korean scientists posting on a blog . . . are credited with actually looking through the details of his work and finding evidence for fabrication.

For whatever reason that the peer review process failed, the Internet-as-peer-reviewer helped to dig out the truth. The Internet will certainly not prevent publication bias, but it is a powerful medium that can communicate almost immediate correction of errors so obvious that they should have been detected before publication.

Involvement of Professional Organizations

It seems that every profession, including science, has its Washington lobby. The American Association for the Advancement of Science (AAAS) sits on prime real estate at 1100 New York Avenue NW and can be thought of as America's science lobby. Do organizations such as that publicize unbalanced or even wrong information about climate change *without* peer review?

On its website, AAAS states, among other activities, that it "spearheads programs that raise the bar of understanding for science worldwide." One such "program" was on June 15, 2004, when AAAS convened what it called an "all-star" panel of U.S. climate scientists to discuss climate change.

The AAAS "all-star" panel including, Daniel Schrag and Michael Oppenheimer, who had appeared less than three weeks earlier onstage with Al Gore during the ultra-Leftist MoveOn.org's kick-off for the climate fiction movie *The Day After Tomorrow*.

Though Schrag and Oppenheimer didn't embrace the nonscience of the movie (e.g., an instantaneous ice age), they did embrace its sentimentality and message, an indication that they believe gross exaggeration and scientifically impossible scenarios are permissible, if they will draw attention to the issue of global climate change.

Other panelists were Sherwood Rowland, Richard Alley, Gerald Meehl, Joyce Penner, and Lonnie Thompson, all prominent in public discussions of climate change.

Reuters' health and science correspondent Maggie Fox noted that the panel expressed frustration that the U.S. government and public were not more concerned with what the panelists saw as the risks associated with global warming.

The panelists were correct in their assertion about public perception of global warming. At that time, a Gallup poll revealed that a plurality of Americans believed that news reports exaggerated the seriousness of global warming. The poll asked this question: "Thinking about what is said in the news, in your view is the seriousness of global warming—generally exaggerated, generally correct, or is it generally underestimated?"

Gallup reported that 38 percent of us thought it is "generally exaggerated," compared with 25 percent who thought it is "generally correct."

Maggie Fox then reported, "[The AAAS panelists] said even as sea levels rise and crop yields fall, officials argue over whether climate change is real and Americans continue to drive fuel-guzzling SUVs."

The statement about crop yields was astoundingly wrong. Figure 7.2 is the history of yields from two important U.S. crops, corn and wheat. There has been a dramatic rise since the late 1940s. The year 2003 saw record-high wheat yields and 2004 saw record-high corn yields. In fact, according to data from the U.S. Department of Agriculture, 14 of the 20 major crops grown here have set record-high yields within the past 10 years. To state that crop yields are falling is at best misleading, and at worse an outright falsehood, given that the U.S. produces almost half of the entire world's corn supply.

Panelist Michael Oppenheimer (Environmental Defense Fund and Princeton University), told the audience: "The sea-level rise over the past century appears greater than what the model says it should be. The [Greenland and Antarctic] ice sheets may be contributing more than the models predict."

Such a statement showed little regard for the latest scientific evidence at that time. For example, published just days before, in the journal *Geophysical Research Letters,* were the results of a sea-level rise study conducted by Cambridge University's Peter Wadhams,

Figure 7.2
U.S. YIELDS OF CORN AND WHEAT, 1900–2007

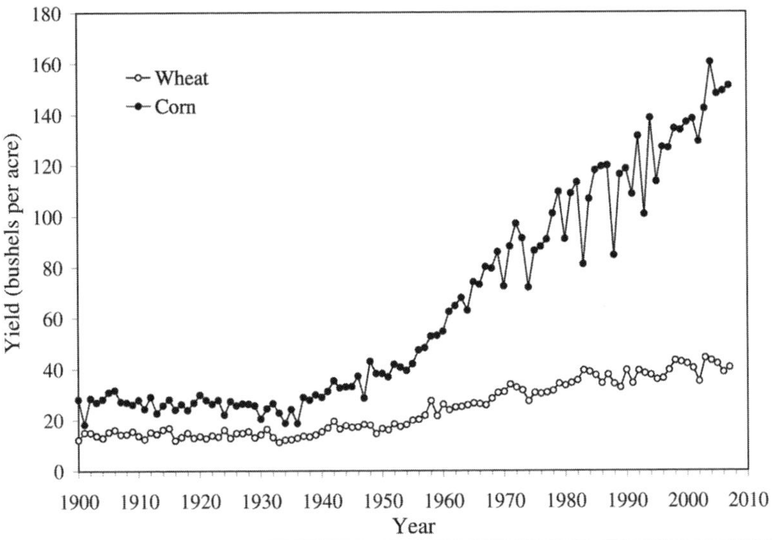

SOURCE: U.S. Department of Agriculture 2008. http://www.nass.usda.gov/QuickStats/.

along with Scripps Institution of Oceanography's Walter Munk. Those researchers carefully calculated the known contributions to sea-level rise (ocean warming, Greenland and Antarctic ice sheets, and midlatitude glaciers) over the 20th century and concluded, "We do obtain a total rise which is at the lower end of the range estimated by the IPCC—exactly the opposite of what Oppenheimer told the AAAS audience."

Publication Bias: Creating the Political Climate

The notion that publication bias is responsible for an unbalanced scientific literature requires that there be some incentive. Obviously, personal and professional advancement are incentives, but what is it that could enable this?

In a word, funding. *Government* funding. Consequently, a thesis of publication bias has to be supported by a notion that the funding stream is in fact predicated upon the belief in dire climatic change, rather than a more moderate view of the subject.

Congress, of course, is the source for federal science funding. How can scientists lead it to a biased point of view? How can powerful scientific organizations such as the IPCC have an influence?

Simple. Make sure that future projections of climate change include a range of estimates, with no particular likelihood ascribed to any value within the range. Any organization that does so can be sure that the most extreme values will be featured in political discussion.

The IPCC is fully aware that its extreme values will be the ones that are quoted for political purposes. In its 2001 assessment, IPCC indicated a prospective 21st century warming of 1.4°C to 5.8°C (2.5°F to 10.4°F). Inevitably, as we have seen, that becomes a statement about warming "as much as 10.4°F."

These high-end estimates are then misused in policy arguments.

Note how the environmental organization Bluewater Network, a part of the radical Friends of the Earth, used extreme values when making a successful political argument.

In March 2004, Bluewater Network took credit for inspiring Senators John McCain (R-AZ) and Ernest Hollings (D-SC) to ask the U.S. General Accounting Office (GAO) to investigate potential impacts of climate change on public lands and waters. Being involved for a long time in environmental politics, Bluewater knew that any such report would be luridly biased.

Bluewater network claimed that GAO did so in response to their 2002 publication *Scorched Earth*, which "examined" the consequences of potential climate change on U.S. public lands and waters. *Scorched Earth* is a typical example of publications from environmental organizations that rely on extreme scenarios and misstated science to suggest that the climate and ecosystems of the United States will be rendered unrecognizable as a result of anthropogenic emissions of greenhouse gases.

The report's Executive Summary contains a paragraph that epitomizes the gloom-and-doom predictions. Given that Senator McCain had become, soon after the GAO report, a principal advocate among Republicans for limits on carbon dioxide emissions, it's worth examining *Scorched Earth*'s assertions in detail.

According to Bluewater,

> Over the past 100 years, emissions of greenhouse gas pollution have led to increased global temperatures of more than

> 1°F, which is unprecedented in the past 1,000 years. Scientists worldwide predict that the pace of global climate change will accelerate over the next century and impact ecosystems with increasingly dramatic results. Average global temperatures could increase by up to 10.4°F, a change unprecedented over the past 10,000 years. This temperature increase is projected to result in reduced water availability, increased catastrophic wildfires and storms, and habitat impacts that could wipe out entire species and ecosystems. Scientists predict a rise in sea level of up to 2.89 feet as a result of projected global temperature increases. Coupled with increasingly severe storm events, a sea-level rise of this magnitude will reshape coastlines and submerge low-elevation islands entirely in both the U.S. and abroad. These global climate change impacts will occur so rapidly that many plant and wildlife species will not survive.

Here's some of their assertions in perspective:

"*. . . increased global temperatures of more than 1°F . . . .*" True, but misleading. Yes, there has been a global temperature rise of about "more than 1°F" in the past 100 years. But about half that rise occurred prior to the mid-1940s—a period before there were large anthropogenic emissions of greenhouse gases. Between the mid-1940s and mid-1970s, global temperatures even declined somewhat, before beginning to rise again from the late 1970s to 1998. A human fingerprint on the more recent temperature increase is probable, because, as predicted by theory, the warming tends to take place in the coldest air of the winter, mainly in Siberia. But it's inaccurate and misleading to state that the temperature rise that began more than 100 years ago has resulted from human-enhanced levels of greenhouse gases in the atmosphere.

"*. . . scientists worldwide predict that the pace of global climate change will accelerate over the next century . . . .*" This results from the coupling of "as much as 10.4°F," which is the extreme value from the IPCC, with the previous statement about the observed rate of warming. Because the IPCC is composed of "scientists worldwide" and the two rates are obviously different, then "scientists worldwide" must predict an increasing rate of warming.

In fact, as shown in Figure 1.5 (see insert), the "consensus" rate of warming for their "midrange" emissions scenario is constant, not increasing. But the coupling of observed and (extreme) forecast warming creates a totally different impression.

"*. . . a rise in sea level of up to 2.89 feet . . . .*" Again, an extreme value is used. The IPCC "Third Assessment Report" (the latest one at the time of the publication of *Scorched Earth*) gives the range of sea-level rise by 2100 resulting from anthropogenic greenhouse gas increases as 0.30 to 2.89 feet. Again, Bluewater emphasizes only the highest end of the range. The low end of this range results in impacts no greater than those we observed during the 20th century, minor changes to which we fully adapted.

"*. . . increasingly severe storm events . . . .*" A common promise, but as we have demonstrated repeatedly in this book, one that is at best controversial, and that certainly is not supported by a thorough reading of the scientific literature.

The Executive Summary of Bluewater Network's *Scorched Earth* is accurate in one respect: It correctly characterizes the misconceptions contained throughout the rest of the report. So much for the publication said to have inspired McCain and Hollings's GAO request.

The GAO's response

Suffice it to say that the GAO took Bluewater's lure hook, line, and sinker, providing one of the most profound examples of publication bias. We'll just excerpt one paragraph, a summary of the biological effects of climate change:

> Biological effects of climate change include increases in insect and disease infestations, shifts in species distribution, coral bleaching, and changes in the timing of natural events, among others. For example, warmer temperatures and reduced precipitation associated with climate change have contributed to insect outbreaks in some areas, as illustrated at the Chugach National Forest in Alaska. According to an FS [Forest Service] official at the forest, a spruce bark beetle outbreak has led to high mortality rates for certain types of spruce trees on over 400,000 acres of the Chugach. In the Kenai Peninsula, Alaska, on which part of the forest is located, about 1 million acres have been affected by the beetles. Officials at the Chugach indicated that continued increases in temperature and decreases in precipitation could further change vegetation composition and structure, and increase the incidence and severity of future insect outbreaks. Similarly, in the Mojave Desert near the BLM Kingman Field

> Office, invasive grasses, combined with drought caused, at least in part, by climate change, have increased the frequency and severity of wildland fires, destroying native plants and transforming some desert communities into annual grasslands. Prolonged drought weakens the natural plant communities and then, in periods of wetness, invasive species—particularly grasses—fill the gaps between native vegetation. These invasive grasses can spread and grow faster than native species; the thicker and less evenly spaced vegetation leads to increased fire danger. If a fire starts, it burns much hotter due to the invasive grasses. Native plant communities, such as saguaro cacti and Joshua trees, are damaged, which provides further environment for invasive species and increased fire danger. According to experts, this shift in ecosystems from desert to grassland is likely to continue as the climate changes, which will in turn result in a loss of species diversity in these areas.

Note, *not one* positive effect of climate change is mentioned. There's no discussion of the fact that carbon dioxide itself has been known to make plants grow better for 100 years.

Consider the work of Ramakrishna Nemani and colleagues, who studied two decades' worth of satellite observations of large-scale plant growth patterns across the world, and published their results in *Science*. They reported a remarkable enhancement of global vegetation during that time. Nemani et al. concluded that the enhanced growth resulted from a combination of two major influences: the increased fertilization effect from growing concentrations of atmospheric carbon dioxide and the patterns of change in the earth's climate during the study period. That finding is strong evidence of a global-scale benefit of uniform greenhouse gas enhancement—and stands in stark contrast to a picture of ecological destruction.

Nor is there any mention of increased crop production as a direct result of carbon dioxide–induced growth stimulation and as an indirect effect of increasing growing seasons and midlatitude precipitation.

What could be a finer example of the workings of publication bias that ultimately influences policy? A lobbying group purposefully ignores anything but the most lurid estimates of the IPCC, then sells their report to two important senators, who commission the GAO to produce a completely one-sided review.

The Hockey Stick Controversy

A prominent example of publication bias via lax reviews in modern climate science is the acceptance, by *Nature*, of a 1,000-year temperature "history" derived from a mathematical combination of large numbers of "paleolclimatic" records—including tree rings, corals, and ice cores, all of which provide some annual information on climate. (As an example, tree rings are thicker in warm, wet summers than they are in cold, dry ones.) It was originally a 600-year record published in *Geophysical Research Letters*, but the methodology for both the 600-year and the more prominent *Nature* paper was the same.

This was discussed briefly earlier in this chapter in the context of reviewer bias as noted in the Wegman report. The work in question is the same "Hockey Stick" (Figure 7.3; see insert), which figured prominently in the widely cited "Third Assessment Report" on climate change published by the IPCC in 2001, and became a poster child for global warming enthusiasts worldwide.

Steve McIntyre, a retired mining executive (and mathematician) from Canada became interested in how this result was obtained and, along with University of Guelph Economist Ross McKitrick, doggedly pursued Mann's original data and then applied the same mathematics to it that Mann had done.

Both were inherently suspicious. In 2005, economist McKitrick wrote, "After the dot-com boom, however, many business people cringe when they see a hockey-stick graph." One of those, of course, was McIntyre.

McIntyre and McKitrick plowed through the mathematical details of the original paper, which McKitrick characterized as "written in grandiose yet disorganized prose and omit[ting] the mathematical equations that would allow readers to attain an unambiguous understanding of what was done."

McIntyre had hypothesized that the method used to reference the data to the 1902–80 period (rather than the 600-year average), combined with the complicated mathematics, would preferentially create hockey sticks.

(continued on next page)

(continued)

The trials and tribulations they went through to obtain the original data and do their analyses is legendary and detailed elsewhere, such as in chapter 2 of the book *Shattered Consensus*, written by Ross McKitrick.

In a section of the *Shattered Consensus* chapter titled "How to Make a Hockey Stick," McKitrick shows the difference between what happens when the mathematics is based on the average of an entire (600-year) sample of random numbers vs. when it is based on the average of the last 78 years (Figure 7.4).

Figure 7.4

MATHEMATICAL SIMULATIONS OF HISTORICAL TEMPERATURES, BASED ON THE AVERAGE OF A 600-YEAR SAMPLE OF RANDOM NUMBERS (TOP); AND THE AVERAGE OF THE PAST 78 YEARS PLUS RANDOM NUMBERS (BOTTOM)

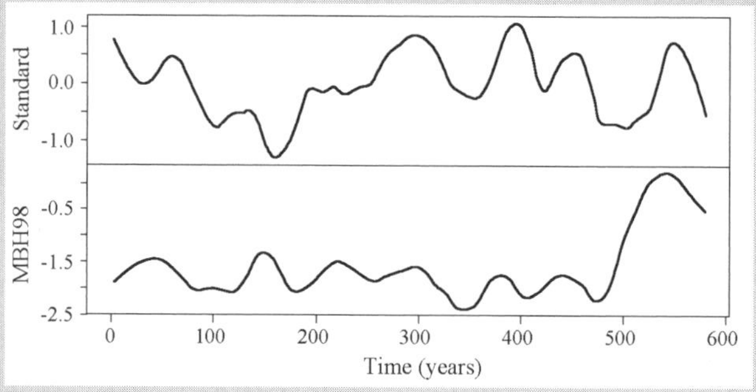

SOURCE: McKitrick 2005.
MBH98 = "Mann, Bradley and Hughes, 1998" method, which used the last 78 years of observed temperatures for calculation of the global temperature index ("hockey stick").

In *Shattered Consensus*, McKitrick blamed the Hockey Stick on lack of rigorous peer review. He and McIntyre learned this first-hand when, after sending their results to *Nature*, the reviewers stated that they were not capable of determining whether Mann or McIntyre and McKitrick were correct. From that, McKitrick concluded:

(continued on next page)

(continued)

> We are quite confident that *Nature*'s peer reviewers for the original publication did not examine the data or the program used to produce Mann's hockey stick or carry out any audit level due diligence.

When the IPCC published Mann's hockey stick prominently in the Technical Summary of the "Third Assesssment Report," there was obviously no attempt to question Mann's findings; they relied only on the fact that it was published in *Nature* and as a result it was obviously beyond question.

Publication bias doesn't mean that *only* alarming global warming papers are published, but rather that they are likely to comprise the majority. Further, it means that the minority of nonalarmist (or anti-alarmist) publications are likely to have undergone a very rigorous review.

The causes are manifold, including the file drawer problem, personal incentives, and the tendency for extreme results to be used for political effect. We are sure that this panoply is only one set of factors that has led to the "climate of extremes," but it has certainly resulted in an ocean of dire findings dotted by islands of moderation.

8. Balancing Act: A Modest Proposal

The last chapter demonstrated that there are substantial institutional biases in climate science that result in a preponderance of gloom-and-doom findings. Yet the world prospers, despite global warming almost always being "worse than we thought."

In the peer review process, there are clear incentives to keep global warming "hot," because the resulting headlines will keep political (and federal funding) attention on it. There are also obvious disincentives to publish much of the other side of the story.

The results are interesting. Horror stories that have received shoddy review are easy to tear up in public. More moderate work that has received stringent review holds up well under scrutiny. Even though the ratio of horror stories to bedtime stories on global warming is around 10 to 1, the more moderate paradigm seems to possess remarkable internal consistency. Put simply, the earth exhibits a modest warming trend; computer models indicate that trend will continue; and that means there is plenty of time—a century or so—for technological development that will be more efficient and emit far less carbon dioxide. Of course, the way to delay that happy ending is to institute policies immediately that are based on the horror stories. All that does is suck investment capital out of the system—capital that could have been used to create a more efficient future.

We see an example of an irrational political response to the horror stories every day now, at the grocery store. President Bush's response to the clamor for global warming policy was to ask for legislation mandating ethanol production. In 2005, a Republican congress sent a Republican president an energy bill that resulted in 7 billion gallons of ethanol in 2007 being produced as a replacement for fossil fuel. The primary feedstock is corn. Put another way, the primary feedstock is *food*. In 2007, we diverted a quarter of our corn crop from food to ethanol.

Never mind that this doesn't do a thing about global warming. As shown in 2008 by Princeton's Timothy Searchinger et al. in *Science* magazine, a study of the entire life cycle of biofuels demonstrates that *they produce more carbon dioxide than they save.* That should surprise no one. Agriculture often requires the cutting down of an original forest. Corn requires fertilizer—a fossil fuel–based product—to achieve high yields. Tractors run on gas or diesel. In the fermentation process, the concentration of ethyl alcohol (ethanol) has to remain below about 20 percent, or the fermenting yeasts will die. So, to create pure ethanol, 80 percent of the liquid has to be boiled away, almost always using fossil fuel–derived heat. Ethanol is a loser—for everyone except agriculture, which benefits enormously from federal subsidies for its production.

In 2007, President Bush proposed that 20 percent of our current gasoline consumption be displaced by ethanol in 2020. If all our corn became ethanol, only 12 percent of our gasoline consumption would be displaced. Meeting this target requires another source of ethanol and new technologies to digest cellulose, making so-called "cellulosic" ethanol. The crop of choice would likely be a tall perennial weed called switchgrass. One problem: cellulosic ethanol has never been produced in an economically viable fashion.

Searchinger noted:

> We found that corn-based ethanol, instead of producing a 20 percent savings [of carbon dioxide], nearly doubles greenhouse emissions over 30 years and increases greenhouse gases for 167 years.

Cellulosic ethanol doesn't work, either. If grown on corn lands, Searchinger calculates that it would raise emissions by 50 percent.

Food prices are skyrocketing. Land that would normally be used to grow soybeans or wheat is diverted to corn for ethanol. Consequently, the price of wheat *tripled* less than three years after the 2005 Energy Bill. Because just about everything (except fresh vegetables) in the grocery store is either a product of primary feedstocks (wheat = bread), or a result of their use (soybean and corn meal = hogs = pork), the price of just about everything is going up.

This creates grumpiness in the United States, and it also fosters riots in poorer places, such as Indonesia and China in early 2008.

That's just one example of how the climate of extremes on global warming can lead to outrageous policy decisions or legal instruments.

Another outrage was the Kyoto Protocol to the UN Framework Convention on Climate Change, a dismal failure. It was supposed to reduce global carbon emissions to about 5 percent below 1990 levels, with the burden falling mainly on the industrialized, developed world (China and India were exempt). Global compliance with Kyoto never happened. Instead, emissions from countries that were supposed to cut emissions went up. The clear reason is that meeting Kyoto was too expensive; if that were not a problem, it would have been implemented easily. Still, even if it were fully complied with, there would be no net detectable reduction in global warming for nearly a century.

So, what can be done to modify the climate of extremes?

Certainly a part of it results from publication bias. If a preponderance of scientific literature says "it's worse than we thought," that will be the consensus position. We demonstrated that there was certainly a bias toward "worse than we thought" articles in *Science* and *Nature*, whatever the cause.

Our modest proposal to reduce publication bias is to eliminate anonymous peer review. The massive expansion of cyberspace makes that possible.

We propose that each of the major journals post on the Web every article that is submitted along with the authors' names. If the journal simply rejected the submission without sending it out for review, that should be noted. If it was sent for review, the journal should post the reviews along with the names of the reviewers.

That would dramatically change peer review. The entire community becomes privy to submitted manuscripts. There's little chance that this plan would lead to plagiarism or theft of concepts because there's a traceable timeline. More important, the wider scientific community can now see what articles are accepted or rejected, and what the critical reviews were.

The net result will be the opening of science. Papers accepted for publication that are highly flawed would probably have undergone very cursory review, which will be duly noted in the science blogosphere. Similarly, those that are rejected without sufficient grounds will also be made known. Remember, it was the blogosphere that first exposed the fraud in the stem cell paper of Woo Suk Hwang discussed in the last chapter.

This is much different from an experiment in "Open Peer Review" that was conducted by *Nature* in 2006. *Nature* asked authors if they would allow anyone to review a submitted manuscript. Reviewers had to identify themselves. On a parallel track, *Nature* also sent each article out for traditional review, where the journal holds the names of the reviewers in confidence. As might be expected, most of the "public" comments were inconsequential, and, in fact, there were surprisingly few in toto.

In an accompanying editorial, ironically, *Nature* noted that the furor caused by *Science*'s publication of the Hwang paper resulted in the creation of an independent review committee to investigate the incident, and that the committee recommended that "journals apply additional scrutiny and risk assessment to papers that are likely to have a significant public impact, such as those with direct implications for policy, public health or climate change."

We think our suggestion would ensure precisely what the committee recommended.

Although publishing signed peer reviews will hardly end publication bias, it certainly might ensure that highly flawed papers that are published because of light reviews are "outed" in public, and further, it should strengthen the review process itself. However, this proposal will *not* stop bias resulting from the file-drawer problem or the sticky nature of paradigm-driven science.

In this book, we attempted to show that there is a consistent body of literature—albiet smaller in size than its counterpart—that argues *for* the existence of climatic change but *against* its more dramatic and apocalyptic interpretations.

The overall picture seems quite clear. Humans are implicated in the planetary warming that began around 1975. Greenhouse gases are likely to be one cause, probably a considerable one, largely because the warming is accentuated at high-latitude land areas in the Northern Hemisphere, and because it is more prevalent in winter than in summer. A stratospheric cooling trend is also consistent with greenhouse warming as well as stratospheric ozone depletion.

Counterfactual is the observation of no net warming (and probably a cooling) of Antarctica, and very conflicting data on Antarctic snowfall, which should increase as a result of warming of the surrounding ocean. Another problem is that there is clearly "nonclimatic" warming in the temperature history, owing to local site and regional and national factors. In general, destitute nations will not make maintaining a high-quality weather or climate network a high priority. It is interesting that when the UN's surface temperature histories are adjusted for this, the frequency distribution of very far above-normal months is modified, and that it much more resembles the satellite data (which are insensitive to local economic or land-use influence). The main change is there are fewer months that are very far above normal.

As far as "iconic" climate change is concerned, the picture is very ambiguous. This is particularly true for hurricanes. The most severe storms in the Atlantic and Western Pacific have become about as frequent as they were in the 1940s to the 1960s, long before the second warming of the 20th century began. Actual observations of hurricane strength and frequency in these two regions are probably reliable back to when "hurricane hunter" aircraft first took to the air. But other "proxy" records of storms, such as cave stalagmites or datable sediments from periodic overwashes, go back hundreds and even thousands of years, indicating nothing unusual in the current regime.

Historical temperatures turn out to be much more problematic than once thought. The three major records (surface thermometers, weather balloons, and satellites) have undergone major revisions, which create "more" warming out of the same initial data. This is like flipping a coin and getting all heads or tails, as presumably it is equally possible that each record would suffer from methodological or technical flaws that would give an equal probability of either raising or lowering the temperature trend when revised. At any rate, the probability for the records to be initially unbiased but to change in one direction for two revisions is 0.016, or less than 1 in 50. What's happened is certainly possible, but it is not very probable.

With regard to the ice and sea-level rise, again we find conflicting data, even though there has been a tremendous amount of press coverage about the demise of Greenland. The most recent decade was certainly no warmer than several in the early 20th century, and long-standing temperature records even hint that it may have been just as warm in the late 18th century. The big to-do about the discovery of "Warming Island" turns out to be a farce. It's shown as an island in a map accompanying a book by aerial photographer Ernst Hofer, published in 1957, near the end of several decades of warm temperatures. Greenland then cooled down and extended an ice bridge to the "island," which was uncovered again in 2005.

Arctic temperatures are going up—and beginning to exceed those observed in the 1930s. Still, there's some pretty strong evidence from the tundra of Siberia and Scandinavia that conditions were much warmer *for millennia* after the end of the last ice age. If the summer sea ice is receded now, it was probably gone then (despite that, the polar bear and the Inuit survived).

Satellite-sensing of sea ice extent began in 1979, when the Arctic was at the end of its coldest period since the early 1920s. Consequently, ice there had to have expanded when that record began, and much of the early decline was merely a return to normal. Since then, late-summer ice has continued to decline because of increasing temperature. But on the same planet, what is to be made of the fact that summer sea-ice extent in the Southern Hemisphere reached record levels in 2007–08, that most "warming" in Alaska is explained as a one-year jump in 1976–77, or that the glaciers of Kilimanjaro were receding when the planet was cooling in the mid-20th century?

The IPCC expresses "low confidence" in any estimate of future temperate-latitude storms. That's probably because, despite a jillion stories in the Euro tabloids, there's no evidence for any long-term trends in storminess there. Here in the States, rainfall is increasing, but the proportion from heavy rainstorms remains the same.

Politicians blame wildfires in Southern California on global warming, with absolutely no supporting evidence from the local climate history. Recent, highly publicized drought in the Pacific Southwest pales when compared with a whopper in the 12th century. Despite a warming trend, satellites show us no global trend whatsoever in fire frequency or extent.

It turns out that Europe's killer heat wave of 2003 was a small atmospheric bubble embedded in a summer known worldwide for its relative moderation. Nonetheless, the more frequent heat waves become, the fewer people die. It's called adaptation, which is physiological as well as political. Consequently, when a similar heat wave hit three years later, there were *fewer* deaths than would have normally been generated by such temperatures. Speaking of a real catastrophe, it turns out that the North Atlantic's circulation has been quite stable, quashing scare stories of an imminent ice age in Europe.

But perhaps the biggest piece of science that has been kept out of public view is the tremendous number of lives that have effectively been saved by the technology powering and developed by our fossil fuel–driven society. When life expectancy doubles, as it has in the industrialized world in the last 110 years, that's equivalent to saving one of every two lives. No one will ever know the number of people who would have otherwise died, but somewhere around a billion of us is a reasonable estimate.

We hope that our readers have enjoyed learning about the real science on global warming that has received so little attention. It paints a compelling picture of a warming planet that steadfastly ignores Cassandra, as people live longer, more prosperous lives.

References and Selected Reading

Note: Following on chapter 7's discussion, we have classified the 116 *Science* and *Nature* references included in our review with the designations (B) for "better" (if the findings reduced an ill effect related to climate change), (W) for "worse" (if the findings increased a negative effect related to climate change), and (N) for "neutral or could not classify" (if the findings were mixed or we were unable to categorize the material). References not included in our review are undesignated.

Ahlmann, H. W. "Researches on Snow and Ice." *The Geographical Journal* 107 (1946): 1,918–40.

______. "The Present Climate Fluctuation." *The Geographical Journal* 112 (1948): 165–93.

______. "Glacier Variations and Climatic Fluctuations." In *Series Three, Bowman Lecture Series, the American Geographical Society*. New York: George Grady Press, 1953 (available from http://www.questia.com/PM.qst?a = o&d = 1918470).

Alley, R. B., et al. "Ice-Sheet and Sea-Level Changes." *Science* 310 (2005): 456–60. (W)

Andreadis, K., and D. Lettenmaier. "Trends in 20th-Century Drought over the Continental United States," *Geophysical Research Letters* 33 (2006): GL025711.

Angell, J., and J. Korshover. "Estimates of the Global Change in Tropospheric Temperature between 1958 and 1973," *Monthly Weather Review* 103 (1975): 1,007–12.

Arctic Council and IASC (International Arctic Science Committee). *Arctic Climate Impact Assessment*. Cambridge: Cambridge University Press, 2005 (available from http://www.acia.uaf.edu/pages/scientific.html).

Balling, R. C. Jr., and S. B. Idso. "Historical Temperature Trends in the United States and the Effect of Urban Population Growth." *Journal of Geophysical Research* 94 (1989): 3,359–63.

Barnett, T. P., J. C. Adam, and D. P. Lettenmaier. "Potential Impacts of a Warming Climate on Water Availability in Snow-Dominated Regions," *Nature* 438 (2005): 303–09. (W)

Barnett, T. P., D. W. Pierce, and R. Schnur. "Detection of Anthropogenic Climate Change in the World's Oceans." *Science* 292 (2001): 270–74.

Barnett, T. P., et al. "Penetration of Human-Induced Warming into the World's Oceans," *Science* 84 (2005): 287. (N)

Barring, L., and H. Von Storch. "Scandinavian Storminess since about 1800." *Geophysical Research Letters* 31 (2004): L20202, doi:10.1029/2004GL020441.

Battisti, D., et al. Brief for Climate Scientists David Battisti et al. as Amici Curiae in Support of Petitioners at 19, *Massachusetts v. U.S. Environmental Protection Agency*, 548 U.S. 903 (2006) (No. 05-1120), 2006 WL 2563377, August 30, 2006.

Bellamy, P. H., et al. "Carbon Losses from All Soils across England and Wales, 1978–2003," *Nature* 437 (2005): 245–48. (W)

Bellouin, N., et al. "Global Estimate of Aerosol Direct Radiative Forcing from Satellite Measurements," *Nature* 438 (2005): 1,138–41. (W)

Bengtsson, L., M. Botzet, and M. Esch. "Will Greenhouse Gas–Induced Warming over the Next 50 Years Lead to a Higher Frequency and Greater Intensity of Hurricanes?" *Tellus* 48A (1996): 57–75.

Bengtsson, L., K. I. Hodges, and E. Roeckner. "Storm Tracks and Climate Change." *Journal of Climate* 19 (2006): 3,518–43.

Beniston, M., and S. Goyette. "Changes in Variability and Persistence of Climate in Switzerland: Exploring 20th-century Observations and 21st-Century Simulations." *Global and Planetary Change* 57 (2007): 1–15.

Benninghoff, W. S. "Interaction of Vegetation and Soil Frost Phenomena," *Arctic* 5 (1952): 34–43.

Biesmeijer, J. C., et al. "Parallel Declines in Pollinators and Insect-Pollinated Plants in Britain and the Netherlands," *Science* 313 (2006): 351–54. (W)

Bindschadler, R. "Climate Change: Hitting the Ice Sheets Where It Hurts," *Science* 311 (2006): 1,720–21. (W)

Blaustein, A. R., and A. Dobson. "Extinctions: A Message from the Frogs," *Nature* 439 (2006): 143–44. (W)

Bluewater Network, 2002. *Scorched Earth: Global Climate Change Impacts on Public Lands and Waters*. http://www.bluewaternetwork.org/reports/rep_ca_global_scorched.pdf.

Both, C., et al. "Climate Change and Population Declines in a Long-Distance Migratory Bird," *Nature* 441 (2006): 81–83. (W)

Boyer, T., et al. "Changes in the Freshwater Content of the Atlantic Ocean, 1955–2006," *Geophysical Research Letters* 34 (2007): L16603, doi:10.1029/2007GL030126.

Bradley, R. S., F. T. Keimig, and H. F. Diaz. "Projected Temperature Changes along the American Cordillera and the Planned GCOS Network," *Geophysical Research Letters* 31 (2004): L16210, doi:10.1029/2004GL020229.

Bradley, R. S., et al. "Threats to Water Supplies in the Tropical Andes," *Science* 312 (2006): 1,755–56. (W)

Bradshaw, W. E., and C. M. Holzapfel. "Evolutionary Response to Rapid Climate Change," *Science* 312 (2006): 1,477–78. (N)

Braganza, K., D. J. Karoly, and J. M. Arblaster. "Diurnal Temperature Range as an Index of Global Climate Change during the Twentieth Century." *Geophysical Research Letters* 31 (2004): L13217, doi:10.1029/2004GL019998.

Briner, J. P., et al. 2006. "A Multi-Proxy Lacustrine Record of Holocene Climate Change on Northeastern Baffin Island, Arctic Canada." *Quaternary Research* 65 (2006): 431–42.

Brohan, P., et al. "Uncertainty Estimates in Regional and Global Observed Temperature Changes: A New Dataset from 1850." *Journal of Geophysical Research* 111 (2006): D12106.

Brommer, D. M., R. S. Cerveny, and R. C. Balling Jr. "Characteristics of Long-Duration Precipitation Events across the United States." *Geophysical Research Letters* 34 (2007): L22712, doi:10.1029/2007GL031808.

Brook, E. J. "Tiny Bubbles Tell All." *Science* 310 (2005): 1,285–87. (N)

Bryden, H. L., H. R. Longworth, and S. Cunningham. "A Slowing of Atlantic Meridional Overturning Circulation at 25°N." *Nature* 438 (2005): 655–57. (W)

Buchanan, J. B., and J. Tullock. *The Calculus of Consent*. Ann Arbor, MI: University of Michigan Press, 1962.

Caias, P., et al. "Europe-Wide Reduction in Primary Productivity Caused by the Heat and Drought in 2003," *Nature* 437 (2005): 529–33. (W)

Cassey, P., et al. "A Survey of Publication Bias within Evolutionary Ecology." *Proceedings: Biological Sciences* 271 (2004): S451–54.
Cazenave, A., "Global Change: Sea Level and Volcanoes," *Nature* 438 (2005): 35–36. (W)
Cess, R. D. "Water Vapor Feedback in Climate Models," *Science* 310 (2005): 795–96. (N)
Chan, A-W., et al. "Empirical Evidence for Selective Reporting of Outcomes in Randomized Trials: Comparison of Protocols to Publications." *Journal of the American Medical Association* 291 (2004): 2,457–65.
Chapin, F. S., III, et al. "Role of Land-Surface Changes in Arctic Summer Warming." *Science* 310 (2005): 657–60. (W)
Chapman, W. L., and J. E. Walsh. "A Synthesis of Antarctic Temperatures." *Journal of Climate* 20 (2007): 4,096–17.
Chase, T. N., et al. "Was the 2003 European Summer Heat Wave Unusual in a Global Context?" *Geophysical Research Letters* 33 (2006): L23709, doi:10.1029/2006GL027470.
Cherry, M. "Ministers Agree to Act on Warnings of Soaring Temperatures in Africa." *Nature* 437 (2005): 1,217. (W)
Christy, J. R., et al. "Analysis of the Merging Procedure for the MSU Daily Temperature Time Series." *Journal of Climate*, 5 (1998): 2,016–41.
Christy, J. R., et. al. "MSU Tropospheric Temperatures: Dataset Construction and Radiosonde Comparisons." *Journal of Atmospheric and Oceanic Technology*, 17(2000): 1,153–170.
Christy, J. R., et al. "Error Estimates of Version 5.0 of MSU-AMSU Bulk Atmospheric Temperatures." *Journal of Atmospheric and Oceanic Technology*, 20 (2003): 613–29.
Church, J. A., N. J. White, and J. M. Arblaster. "Significant Decadal-Scale Impact of Volcanic Eruptions on Sea Level and Ocean Heat Content." *Nature* 438 (2005): 74–77. (W)
Chylek, P., et al. "Greenland Warming of 1920–1930 and 1995–2005." *Geophysical Research Letters* 33 (2006): L11707, doi:10.1029/2006GL026510.
Chylek, P., et al. "Remote Sensing of Greenland Ice Sheet Using Multispectral Near-Infrared and Visible Radiation." *Journal of Geophysical Research.* 112 (2007): D24S20, doi: 10.1029/2007/JD008742,2007.
Coakley, J. "Atmospheric Physics: Reflections on Aerosol Cooling." *Nature* 438 (2005): 1,091–92. (W)
Covey, C., et al. "An Overview of Results from the Coupled Model Intercomparison Project." *Global And Planetary Change* 37 (2003): 103–33.
Cryosphere Today, 2007. http://arctic.atmos.uiuc.edu/cryosphere/.
Cubasch, U., et al. "Projections of Future Climate Change." In *Climate Change 2001: The Scientific Basis,* edited by J. T. Houghton et al., 525–82. Cambridge, U.K.: Cambridge University Press, 2001.
Cullen, N. J., et al. "Kilimanjaro Glaciers: Recent Areal Extent from Satellite Data and New Interpretation of Observed 20th-Century Retreat Rates." *Geophysical Research Letters* 33 (2006): L16502, doi:10.1029/2006GL027084.
Curry, R., et al. "A Change in the Freshwater Balance of the Atlantic Ocean over the Past Four Decades." *Nature* 426 (2003): 826–29.
Danish Meteorological Institute, 2008. http://www.dmi.dk/dmi/tr08-04.pdf.
Davey, C. A., and R. A. Pielke Sr. "Microclimate Exposures of Surface-based Weather Stations: Implications for the Assessment of Long-Term Temperature Trends." *Bulletin of the American Meterological Society* 86 (2005): 497–504.

Davidson, E. A., and I. A. Janssens. "Temperature Sensitivity of Soil Carbon Decomposition and Feedbacks to Climate Change." *Nature* 440 (2006): 165–73. (W)

Davis, C. H., et al. "Snowfall-Driven Growth in East Antarctic Ice Sheet Mitigates Recent Sea-Level Rise." *Science* 308 (2005): 1,898–1,901.

Davis, R. E., et al. "Changing Heat-Related Mortality in the United States." *Environmental Health Perspectives* 14 (2003): 1,712–18.

Davis, R. E., et al. "A Climatology of Snowfall–Temperature Relationships in Canada." *Journal of Geophysical Research* 104 (1999): 11,985–94.

De Angelis, C., et al. "Clinical Trial Registration: A Statement from the International Committee of Medical Journal Editors." *Medical Journal of Australia* 181 (2004): 293–94.

De Laat, A. T. J., and A. N. Maurellis. "Industrial CO_2 Emissions as a Proxy for Anthropogenic Influence on Lower Tropospheric Temperature Trends." *Geophysical Research Letters* 31 (2004): L05204, doi:10.1029/2003GL019024.

De Wit, M., and J. Stankiewicz, "Changes in Surface Water Supply across Africa with Predicted Climate Change." *Science* 311 (2006): 1,917–21. (W)

Delisle, G. "Near-Surface Permafrost Degradation: How Severe during the 21st Century?" *Geophysical Research Letters* 34 (2007): L09503, doi:10.1029/2007GL029323.

Domack, E., et al. "Stability of the Larsen B Ice Shelf on the Antarctic Peninsula during the Holocene Epoch." *Nature* 436 (2005): 681–85. (W)

Donnelly, J. P., and J. D. Woodruff. "Intense Hurricane Activity over the Past 5,000 Years Controlled by El Niño and the West African Monsoon." *Nature* 447 (2007): 465–68.

Doran, P. T., et al. "Antarctic Climate Cooling and Terrestrial Ecosystem Response." *Nature* 415 (2002): 517–20.

Douglass, D. H., et al. "A Comparison of Tropical Temperature Trends with Model Predictions." *International Journal of Climatology* (2007): DOI: 10.1002/joc.1651.

Dowdeswell, J. A. "Atmospheric Science: The Greenland Ice Sheet and Global Sea-Level Rise." *Science* 311 (2006): 963–64. (W)

Duplessy, J. C., et al. "Holocene Paleoceanography of the Northern Barents Sea and Variations of the Northward Heat Transport by the Atlantic Ocean." *Boreas* 30 (2001): 2–16.

Ekstrom, G., M. Nettles, and V. C. Tsai. "Seasonality and Increasing Frequency of Greenland Glacial Earthquakes," *Science* 311 (2006): 963–64. (W)

Emanuel, K. "Increasing Destructiveness of Tropical Cyclones over the Past 30 Years." *Nature* 436 (2005a): 686–88. (W)

———. "Meteorology: Emanuel Replies." *Nature* 438 (2005b): E12. (W)

Environment America, 2007. http://www.environmentamerica.org/home/reports/report-archives/global-warming-solutions/global-warming-solutions/when-it-rains-it-pours-global-warming-and-the-rising-frequency-of-extreme-precipitation-in-the-united-states.

Evans, M. N. "Paleoclimatology: The Woods Fill up with Snow." *Nature* 440 (2006): 1,120–21. (N)

Feddema, J. J., et al. "The Importance of Land-Cover Change in Simulating Future Climates." *Science* 310 (2005): 1,674–78. (W)

Federov, A. V., et al. "The Pliocene Paradox (Mechanisms for a Permanent El Niño)." *Science* 312 (2006): 1,485–89. (W)

Field, D. B., et al. "Planktonic Formanifera of the California Current Reflect 20th-Century Warming." *Science* 311 (2006): 63–66. (N)

Foley, J. A. "Atmospheric Science: Tipping Points in the Tundra." *Science* 310 (2005): 627–28. (W)

Fouillet, A., et al. "Has the Impact of Heat Waves on Mortality Changed in France since the European Heat Wave of Summer 2003? A Study of the 2006 Heat Wave." *International Journal of Epidemiology* (2008): doi:10.1093/ije/dym253.

Free, M., et al. "Radiosonde Atmospheric Temperature Products for Assessing Climate (RATPAC): A New Set of Large-Area Anomaly Time Series." *Journal of Geophysical Research* 110 (2005): D22101, doi: 10.1029/2005JD006169.

Garrett, T. J., and C. Zhao. "Increased Arctic Cloud Longwave Emissivity Associated with Pollution from Mid-Latitudes." *Nature* 440 (2006): 787–89. (W)

Gedney, N., et al. "Detection of a Direct Carbon Dioxide Effect in Continental River Runoff Records." *Nature* 439 (2006): 835–36. (N)

Gerdel, R. W. "A Climatological Study of the Greenland Ice Sheet." *Folia Geographia Danica* IX (1961).

Gillett, N. P. "Climate Modeling: Northern Hemisphere Circulation." *Nature* 437 (2005): 496. (W)

Girardin, M. P., J. Tardif, and M. D. Flannigan. "Temporal Variability in Area Burned for the Province of Ontario, Canada, during the Past 200 Years Inferred from Tree Rings." *Journal of Geophysical Research* 111 (2006): D17108, doi:10.1029/2005JD006815.h

Gleckler, P. J., et al. "Volcanoes and Climate: Krakatoa's Signature Persists in the Ocean." *Nature* 439 (2006): 675. (W)

Goodridge, J. D. "Urban Bias Influences on Long-Term California Air Temperature Trends." *Atmospheric Environment* 26B (1992): 1–7.

Gore, A. *An Inconvenient Truth.* Rodale, PA: Rodale Press, 2006.

Gould, S. J. *The Structure of Evolutionary Theory.* Cambridge, MA: Belknap Press, 2002.

Grebmeier, J. M., et al. "A Major Ecosystem Shift in the Northern Bering Sea." *Science* 311 (2006): 1,461–64. (N)

Groode, T. A., and J. B. Haywood, *Biomass to Ethanol: Potential Production and Environmental Impacts.* Cambridge, MA: Massachusetts Institute of Technology Laboratory for Energy and the Environment, 2008.

Hamilton, R. A., et al. "British North Greenland Expedition 1952–4: Scientific Results." *The Geographical Journal* 122 (1956): 203–37.

Hansen, J. E. "The Global Warming Time Bomb?" *Natural Science* (2003). http://naturalscience.com/ns/articles/01-16/ns_jeh.html.

______. "Scientific Reticence and Sea Level Rise." *Environmental Research Letters* (2007): 2 024002, doi: 10.1088/1748-9326/2/2/024002.

Hansen, J. E., and M. Sato. "Trends of Measured Climate Forcing Agents." *Proceedings of the National Academy of Sciences* 98 (2001): 14,778–83.

Harper, J. R. "Coastal Erosion Rates along the Chukchi Sea Coast near Barrow Alaska." *Arctic* 31 (1978): 428–33.

Hartmann, B., and G. Wendler. "The Significance of the 1976 Pacific Climate Shift in the Climatology of Alaska." *Journal of Climate* 18 (2005): 4,824–39.

Hartwell, A. D. "Classification and Relief Characteristics of Northern Alaska's Coastal Zone." *Arctic* 26 (1973): 244–52.

Hasegawa, A., and S. Emori. "Tropical Cyclones and Associated Precipitation over the Western North Pacific: T106 Atmospheric GCM Simulation for Present-Day and Doubled CO_2 Climates." *Scientific Online Letters on the Atmosphere* 1 (2006): 145–48.

Hatun, H., et al. "Influence of the Atlantic Subpolar Gyre on the Thermohaline Circulation." *Science* 309 (2005): 1,841–44. (B)

Heath, J., et al. "Rising Atmospheric CO_2 Reduces Sequestration of Root-Derived Soil Carbon." *Science* 309 (2005): 1,711–13. (W)

Hegerl, G. C., and N. L. Bindoff. "Warming the World's Oceans." *Science* 309 (2005): 254–55. (N)

Hegerl, G. C., et al. "Climate Sensitivity Constrained by Temperature Reconstructions over the Past Seven Centuries." *Nature* 440 (2006): 1,029–32. (B)

Hofer, E. *Arctic Riviera.* Bern, Switzerland: Kümmerly & Frey, Geographical Publishers, 1957.

Hopkin, M. "Antarctic Ice Puts Climate Predictions to the Test." *Nature* 438 (2005): 536–37. (W)

———. "Biodiversity and Climate Form Focus of Forest Canopy Plan." *Nature* 436 (2005): 452. (W)

Howat, I. M., et al. "Rapid Changes in Ice Discharge from Greenland Outlet Glaciers." *Science* 315 (2007): 1,559–61.

Hoyos, C. D., et al. "Deconvolution of the Factors Contributing to the Increase in Global Hurricane Intensity." *Science* 312 (2006): 94–97.

Hu, F.S., et at. "Pronounced Climatic Variations in Alaska During the Last Two Millennia," *Proc. Natl. Acad. Sci.* 98 (2001): 10552-56.

Hume, J. D., and M. Schalk. "Shoreline Processes near Barrow Alaska: A Comparison of the Normal and the Catastrophic." *Arctic* 20 (1967): 86–103.

Hume, J. D., et al. "Short-Term Climate Changes and Coastal Erosion, Barrow Alaska." *Arctic* 25 (1972): 272–78.

Hwang, H. S., et al. "Evidence of a Pluripotent Human Embryonic Stem Cell Line Derived from a Cloned Blastocyst." *Science* 303 (2004): 1,669–74.

Ibisch, P. L., M. D. Jennings, and S. Kreft. "Biodiversity Needs the Help of Global Change Managers, not Museum-Keepers." *Nature* 438 (2005): 156. (W)

Intergovernmental Panel on Climate Change. *Climate Change 2001: The Scientific Basis.* Cambridge, U.K.: Cambridge University Press, 2001.

———. *Climate Change 2007: The Physical Science Basis.* S. Solomon et al. (eds.). Cambridge, U.K.: Cambridge University Press, 2007.

———. *Climate Change 2007: Synthesis Report.* Geneva, Switzerland: Intergovernmental Panel on Climate Change, 2007.

Jackson, R. B., et al. "Trading Water for Carbon with Biological Carbon Sequestration." *Science* 310 (2005): 1,944–47. (W)

Johannessen, O. M., et al. "Arctic Climate Change: Observed and Modeled Temperature and Sea-Ice Variability." *Tellus* 56A (2004): 328–41.

Johnsen, S. J., et al. "Oxygen Isotope and Palaeotemperature Records from Six Greenland Ice-Core Stations: Camp Century, Dye-3, GRIP, GISP2, Renland, and North-GRIP." *Journal of Quaternary Science* 16 (2001): 299–307.

Jones, J. M., and M. Widmann. "Early Peak in Antarctic Oscillation Index." *Nature* 432 (2004): 290–91.

Jones, K. *Sharpening the Focus on Climate Change in the Pacific Northwest.* Seattle, WA: Washington Public Policy Center, 2007.

Jonzen, N., et al. "Rapid Advance of Spring Arrival Dates in Long-Distance Migratory Birds." *Science* 312 (2006): 1,959–61. (N)

Joughin, I. "Climate Change: Greenland Rumbles Louder as Glaciers Accelerate." *Science* 311 (2006): 1,719–20. (W)

Kalnay, E., and M. Cai. "Impacts of Urbanization and Land Use Change on Climate." *Nature* 423 (2003): 528–31.

Karl, T. R., and R. W. Knight. "Secular Trends in Precipitation Amount, Frequency, and Intensity in the United States." *Bulletin of the American Meteorological Society* 79 (1998): 231–241.

Kaser, G., et al. "Modern Glacial Retreat on Kilimanjaro as Evidence of Climate Change: Observations and Facts." *International Journal of Climatology.* 24 (2004): 329–39.

Kaufman, D. S., et al. "Holocene Thermal Maximum in the Western Arctic (0–180°W)." *Quaternary Science Reviews* 23 (2004): 529–60.

Keenlyside, N. S., et al. "Advancing Decadal-Scale Climate Prediction in the North Atlantic Sector." *Nature* 453 (2008): 84–88.

Keim, B. D., et al. "Are There Spurious Temperature Trends in the United States Climate Division database?" *Geophysical Research Letters* 30 (2003): doi:10.1029/202GL016295.

Keppler, F., et al. "Methane Emissions from Terrestrial Plants under Aerobic Conditions." *Nature* 439 (2006): 187–91. (W)

King, A. W., et al. "Plant Respiration in a Warmer World." *Science* 312 (2006): 536–37. (B)

Kitzberger, T., et al. "Contingent Pacific-Atlantic Ocean Influence on Multicentury Wildfire Synchrony over Western North America." *Proceedings of the National Academy of Sciences* 104 (2007): 543–48.

Klein, W. H., and H. R. Glahn. "Forecasting Weather by Means of Model Output Statistics." *Bulletin of the American Meteorological Society* 55 (1974): 1,217–27.

Klotzbach, P .J. "Trends in Global Tropical Cyclone Activity over the Past Twenty Years (1986–2005)." *Geophysical Research Letters* 33 (2006): L010805, doi:10.1029/2006GL025881.

Klotzbach, P. J., and W. M. Gray. "Causes of the Unusually Destructive 2004 Atlantic Basin Hurricane Season." *Bulletin of the American Meteorological Society* 87 (2006): 1,325–33.

Knappenberger, P. C., P. J. Michaels, and R. E. Davis. "The Nature of Observed Climate Changes across the United States during the 20th Century." *Climate Research* 17 (2001): 45–53.

Knight, J. R., et al. "A Signal of Natural Thermohaline Cycles in Observed Climate." *Geophysical Research Letters* (2005): L20708, doi:10.1029/2005GL24233.

Knutson, T. R., and R. E. Tuleya. "Impact of CO_2-Induced Warming on Simulated Hurricane Intensity and Precipitation: Sensitivity to the Choice of Climate Model and Convective Parameterization." *Journal of Climate* (2004): 3,477–95.

Koricheva, J. "Non-Significant Results in Ecology: A Burden or a Blessing in Disguise?" *Oikos* 102 (2003): 397–401.

Korner, C., et al. "Carbon Flux and Growth in Mature Deciduous Forest Trees." *Science* 309 (2005): 1,360–62. (W)

Krabill, W., et. al. "Greenland Ice Sheet: High Elevation Balance and Peripheral Thinning." *Science* 289 (2000): 428–30.

Kuhn, T. S. *The Structure of Scientific Revolutions.* Chicago: University of Chicago Press, 1962.

Kunkel, K. Climate Specialty Group plenary session, Association of American Geographers Climate Day, Chicago, IL, March 8, 2006.

Laaidi, M., K. Laaidi, and J.-P. Besancenot. "Temperature-Related Mortality in France, a Comparison Between Regions with Different Climates from the Perspective of Global Warming." *International Journal of Biometeorology* 51 (2006): 145–53.

Landsea, C. W. "Hurricanes and Global Warming." *Nature* 438 (2005): E11–12. (N)

Landsea, C. W., et al. "Can We Detect Trends in Extreme Tropical Cyclones?" *Science* 313 (2006): 452–54. (N)

Lanzante, J. R., S. A. Klein, and D. J. Seidel. "Temporal Homogenization of Monthly Radiosonde Temperature Data. Part I: Methodology." *Journal of Climate* 16 (2003): 224–40.

Lemke, P., et al. "Observations: Changes in Snow, Ice and Frozen Ground," In *Climate Change 2007: The Physical Science Basis. Contribution of Working Group I to the Fourth Assessment Report of the Intergovernmental Panel on Climate Change,* edited by S. D. Solomon et al., 337–83. Cambridge, U.K.: Cambridge University Press, 2007.

Lewellen, R. "A Study of Beaufort Sea Coastal Erosion, Northern Alaska. Environmental Assessment of the Alaskan Continental Shelf." *U.S. National Oceanic and Atmospheric Administration Annual Report* (1977): 491–527.

Levitus, S., et al. "Warming of the World Ocean." *Science* 287 (2000): 2,225–28.

Liu, J. L.Y., and G. Altman. "Conducting and Reporting of Clinical Trials." *Science* 308 (2005): 201–02.

Long, S. P., et al. "Food for Thought: Lower-Than-Expected Crop Yield Stimulation with Rising CO_2 Concentrations." *Science* 312 (2006): 1,918–21. (W)

Lowe, D. C. "Global Change: A Green Source of Surprise." *Nature* 439 (2006): 148–49. (W)

Lubin, D., and A. M. Vogelmann. "A Climatologically Significant Aerosol Longwave Indirect Effect in the Arctic." *Nature* 439 (2006): 453–56. (W)

Luthcke, S. B., et al. "Recent Greenland Ice Mass Loss by Drainage System from Satellite Gravity Observations." *Science* 314 (2006): 1,286–89.

MacCarthy, G. R. "Recent Changes in the Shoreline near Point Barrow, Alaska." *Arctic* 6(1) (1953): 44–51.

MacDonald, G. M., et al. "Holocene Treeline History and Climate Change across Northern Eurasia." *Quaternary Research* 53 (2000): 302–11.

Mann, M. E., R. S. Bradley, and M. K. Hughes. "Global-Scale Temperature Patterns and Climate Forcing over the Past Six Centuries." *Nature* 392 (1998): 779–87.

———. "Northern Hemisphere Temperatures during the Past Millennium: Inferences, Uncertainties, and Limitations." *Geophysical Research Letters* 26 (1999): 759–62.

Manning, M. R., et al. "Short-Term Variations in the Oxidizing Power of the Atmosphere." *Nature* 436 (2005): 1,001–04. (N)

Masiokas, M. H., et al. "Snowpack Variations in the Central Andes of Argentina and Chile, 1951–2005: Large-Scale Atmospheric Influences and Implications for Water Resources in the Region." *Journal of Climate* 19 (2006): 6,334–52.

Matthews, D. "The Water Cycle Freshens Up." *Nature* 439 (2006): doi: 10.1038/439793a (N)

Matulla, C., et al. "European Storminess: Late Nineteenth Century to Present." *Climate Dynamics* (2007): DOI 10.1007/s00382-007-0333-y.

McKitrick, R. R. "The Mann et al. Northern Hemisphere 'Hockey Stick' Index: A Tale of Due Diligence." In *Shattered Consensus*, edited by P. J. Michaels, 22–49. New York: Rowman and Littlefield, 2005.

McKitrick, R. R., and P. J. Michaels. "Quantifying the Influence of Anthropogenic Surface Processes and Inhomogeneities on Gridded Global Climate Data." *Journal of Geophysical Research* 112 (2007): D24S09, doi: 10.1029/2007JD008465.

Mears, C. A., and F. J. Wentz. "The Effect of Diurnal Correction on Satellite-Derived Lower Tropospheric Temperature." *Science* 309 (2005): 1,548–51. (W)
Meehl, G. A., et al. "The Coupled Model Intercomparison Project (CMIP)." *Bulletin of the American Meteorological Society* 81 (2000): 313–18.
Meko, D., et al. "Medieval Drought in the Upper Colorado River Basin." *Geophysical Research Letters* 34 (2007): L10705, doi: 10.1029/2007GL029988.
Michaels, P. J. *Meltdown: The Predictable Distortion of Global Warming by Scientists, Politicians, and the Media.* Washington: Cato Institute, 2004.
Michaels, P. J. (ed.). *Shattered Consensus: The True State of Global Warming.* New York: Rowman and Littlefield, 2005.
Michaels, P. J., and R. C. Balling Jr. *The Satanic Gases.* Washington: Cato Institute, 2000a.
Michaels, P. J., and P. C. Knappenberger. "Natural Signals in the MSU Lower Tropospheric Temperature Record." *Geophysical Research Letters* 27 (2000b): 2,905–8.
Michaels, P. J., P. C. Knappenberger, and C. W. Landsea. "Extended Comment on: 'Impacts of CO_2-Induced Warming on Simulated Hurricane Intensity and Precipitation: Sensitivity to the Choice of Model and Convective Scheme.' " *Journal of Climate* 18 (2005): 5,179–82.
Michaels, P. J., et al. "Revised 21st-Century Temperature Projections." *Climate Research* 23 (2001): 1–9.
Michaels, P. J., et al. "Trends in Precipitation on the Wettest Days of the Year across the Contiguous United States." *International Journal of Climatology* 24 (2004): 1,872–82.
Michaels, P. J., et al. "Sea-Surface Temperatures and Tropical Cyclones in the Atlantic Basin." *Geophysical Research Letters* 33 (2006): doi: 10.1029/2006GL025757.
Miller, K. G., et al. "The Phanerozoic Record of Global Sea-Level Change." *Science* 310 (2005): 1,293–98. (B)
Milly, P. C. D., K. A. Dunne, and A. V. Vecchia. "Global Pattern of Trends in Streamflow and Water Availability in a Changing Climate." *Nature* 438 (2005): 347–50. (W)
Mitchell, J. M., Jr. "On the Causes of Instrumentally-Observed Secular Temperature Trends." *J. Meteor. 10 (1953); 244–61.*
Mollicone, D., H. D. Eva, and F. Achard. "Ecology: Human Role in Russian Wild Fires." Nature 440 (2006): 436–37. (B)
Monaghan, A. J., et al. "Insignificant Change in Antarctic Snowfall since the International Geophysical Year." *Science,* 313 (2006): 827–831, doi:10.1126.
Monson, R. K., et al. "Winter Forest Soil Respiration Controlled by Climate and Microbial Community Composition." *Nature* 439 (2006): 711–14. (B)
Moran, K., et al. "The Cenozoic Paleoenvironment of the Arctic Ocean." *Nature* 441 (2006): 601–05. (N)
Muscheler, R., et al. "Climate: How Unusual is Today's Solar Activity?" *Nature* 436 (2005): E3–E4. (W)
NASA, 2007. http://www.nasa.gov/vision/earth/environment/greenland_recordhigh.html.
National Climatic Data Center, 2008, http://www.ncdc.noaa.gov/oa/climate/research/ushcn/.
National Climatic Data Center, 2008, http://www7.ncdc.noaa.gov/CDO/CDODivisionalSelect.jsp.
National Research Council. *Reconciling Observations of Global Temperature Change.* Washington, DC: National Academies Press, 2000.
______. *Abrupt Climate Change: Inevitable Surprises.* Washington, DC: National Academies Press, 2001.

———. *Climate Change Science: An Analysis of Some Key Questions.* Washington, DC: National Academies Press, 2001.

Nemani, R. R., et al. "Climate-Driven Increases in Global Terrestrial Net Primary Production from 1982 to 1999." *Science* 300 (2003): 1,560–63.

Nott, J., et al. "Greater Frequency Variability of Landfalling Tropical Cyclones at Centennial Compared to Seasonal and Decadal Scales." *Earth and Planetary Science Letters* 255 (2007): 367–72.

Nunes, F., and R. D. Norris. "Abrupt Reversal in Ocean Overturning during the Palaeocene/Eocene Warm Period." *Nature* 439 (2006): 60–63. (W)

Nyberg, J., et al. "Low Atlantic Hurricane Activity in the 1970s and 1980s Compared to the Past 270 Years." *Nature* 447 (2007): 698–702.

Orr, J. C., et al. "Anthropogenic Ocean Acidification over the Twenty-First Century and its Impact on Calcifying Organisms." *Nature* 437 (2005): 681–86. (W)

Otto-Bliesner, B. L., et al. "Simulating Arctic Climate Warmth and Icefield Retreat in the Last Interglaciation." *Science* 311 (2006): 1,751–53. (W)

Overland, J. E., and K. Wood. "Accounts from 19th-Century Canadian Arctic Explorers' Logs Reflect Present Climate Conditions." *EOS Transactions of the American Geophysical Union* 84 (2003).

Overpeck, J. T., et al. "Paleoclimatic Evidence for Future Ice-Sheet Instability and Rapid Sea-Level Rise." *Science* 311 (2006): 1,747–50. (W)

Palter, J. B., M. S. Lozier, and R. T. Barber. "The Effect of Advection on the Nutrient Reservoir in the North Atlantic Subtropical Gyre." *Nature* 347 (2005): 687–92. (B)

Patel, S. S. "Climate Science: A Sinking Feeling." *Nature* 440 (2006): 734–36. (N)

Patz, J. A., et al. "Impact of Regional Climate Change on Human Health." *Nature* 438 (2005): 310–17. (W)

Pelejero, C., et al. "Preindustrial to Modern Interdecadal Variability in Coral Reef pH." *Science* 309 (2005): 2,204–07. (W)

Pielke, R. A., Jr. "Are There Trends in Hurricane Destruction?" *Nature* 438 (2005): E1.1 (N)

Pielke R. A., Jr., et al. "Hurricanes and Global Warming." *Bulletin of the American Meteorological Society* 86 (2005): 1,571–75.

Pielke, R. A., Sr. "Atmospheric Science: Land Use and Climate Change." *Science* 310 (2005): 1,625–26. (W)

Porter, J. R. "Rising Temperatures Are Likely to Reduce Crop Yields." *Nature* 346 (2005): 174. (W)

Pounds, J. A., et al. "Widespread Amphibian Extinctions from Epidemic Disease Driven by Global Warming." *Nature* 439 (2006): 161–67. (W)

Quadfasel, D. "Oceanography: The Atlantic Heat Conveyor Slows." *Nature* 438 (2005): 565–66. (W)

Ramaswamy, V., et al. "Anthropogenic and Natural Influences in the Evolution of Lower Stratospheric Cooling." *Science* 311 (2006): 1,138–41. (N)

Raper, S. C. B., and R. J. Braithwaite. "Low Sea Level Rise Projections from Mountain Glaciers and Icecaps under Global Warming." *Nature* 439 (2006): 311–13.

Riaño, D., et al. "Global Spatial Patterns and Temporal Trends of Burned Area between 1981 and 2000 Using NOAA-NASA Pathfinder." *Global Change Biology* 13 (2007): 40–50.

Rignot, E., and P. Kanagaratnam. "Changes in the Velocity Structure of the Greenland Ice Sheet." *Science* 311 (2006): 986–90. (W)

Rignot, E., et al. "Recent Antarctica Ice Mass Loss from Radar Interferometry and Regional Climate Modeling." *Nature Geoscience* (2007): doi: 10.1038/neo102.

Robeson, S. "Relationships between Mean and Standard Deviation of Air Temperature: Implication for Global Warming." *Climate Research* 22 (2002): 205–13.

Rosenfeld, D. "Atmosphere: Aerosols, Clouds, and Climate." *Science* 312 (2006): 1,323–24. (W)

Rosenthal, R. "The 'File Drawer Problem' and Tolerance for Null Results." *Psychological Bulletin,* 86 (1979): 638–41.

Salwitch, R. J. "Atmospheric Chemistry: Biogenic Bromine." *Nature* 439 (2006): 275–77. (W)

Santer, B. D., et al. "Amplification of Surface Temperature Trends and Variability in the Tropical Atmosphere." *Science* 309 (2005): 1,551–56. (W)

Scafetta, N., and B. West. "Phenomenological Solar Contribution to the 1900–2000 Global Surface Warming." *Geophysical Research Letters* (2006): doi: 1029/2005GL025539.

______. "Phenomenological Reconstructions of the Solar Signature in the Northern Hemisphere Surface Temperature Records since 1600." *Journal of Geophysical Research* (2007): doi: 10.1029/2007JD008437.

Schiermeier, Q. "Natural Disasters: The Chaos to Come." *Nature* 438 (2005): 903–06. (W)

______. "Arctic Ecology: On Thin Ice." *Nature* 441 (2006): 146–47. (W)

______. "Climate Change: A Sea Change." *Nature* 439 (2006): 256–60. (W)

______. "Insurers' Disaster Files Suggest Climate Is Culprit." *Nature* 411 (2006): 674–75. (W)

______. "Methane Finding Baffles Scientists." *Nature* 439 (2006): 128. (W)

Schimel, D. "Climate Change and Crop Yields: Beyond Cassandra." *Science* 312 (2006): 1,889–90. (W)

Schroter, D., et al. "Ecosystem Service Supply and Vulnerability to Global Change in Europe." *Science* 310 (2005): 1,333–37. (W)

Schulze, E. D., and A. Freibauer. "Environmental Science: Carbon Unlocked from Soils." *Nature* 437 (2005): 205–06. (W)

Schwartz, P., and D. Randall. *An Abrupt Climate Change Scenario and Its Implications for United States National Security.* Report for the U.S. Department of Defense produced by the Global Business Network, 2003, 22.

Schar, C., et al. "The Role of Increasing Temperature Variability in European Summer Heatwaves." *Nature* 427 (2004): 332–36.

Scileppi, E., and J. P. Donnelly. "Sedimentary Evidence of Hurricane Strikes in Western Long Island, New York." *Geochemistry, Geophysics, Geosystems* 8(6) (2007): 1–25.

Searchinger, T., et al. "Use of U.S. Croplands for Biofuels Increases Greenhouse Gases Through Emissions from Land-Use Change." *Science* 319 (2008): 1,238–40.

Semenov, V. A., and L. Bengtsson. "Secular Trends in Daily Precipitation Characteristics: Greenhouse Gas Simulation with a Coupled AOGCM." *Climate Dynamics* 19 (2002): 123–40.

______. "Modes of the Wintertime Arctic Air Temperature Variability." *Geophysical Research Letters* 30 (2003): 1,781–84.

Serreze, M. C., et al. "Observational Evidence of Recent Change in the Northern High-Latitude Environment." *Climatic Change* 46 (2000): 159–207.

Shepherd, A., and D. J. Wingham. "Recent Sea-Level Contributions of the Antarctic and Greenland Ice Sheets." *Science* 315 (2007): 1,529–32.

Sherwood, S. C., J. R. Lanzante, and C. L. Meyer, "Radiosonde Daytime Biases and Late 20th-Century Warming." *Science* 309 (2005): 1,556–59. (W)

Shindell, D. T., and G. A. Schmidt. "Southern Hemisphere Climate Response to Ozone Changes and Greenhouse Gas Increases." *Geophysical Research Letters* 31 (2004): L18209, doi:10.1029/2004GL020724.

Siegenthaler, U., et al. "Stable Carbon Cycle–Climate Relationship during the Late Pleistocene." *Science* 310 (2005): 1,313–17. (N)

Silliman, B. R., et al. "Drought, Snails, and Large-Scale Die-Off of Southern U.S. Salt Marshes." *Science* 310 (2005): 1,803–06.

Slujis, A., et al. "Subtropical Arctic Ocean Temperatures during the Palaeocene/Eocene Thermal Maximum." *Nature* 441 (2006): 610–13. (W)

Soden, B. J., et al. "The Radiative Signature of Upper Tropospheric Moistening." *Science* 310 (2005): 841–44. (N)

Sowers, T. "Late Quaternary Atmospheric CH4 Isotope Record Suggests Marine Clathrates Are Stable." *Science* 311 (2006): 838–40. (B)

Spahni, R., et al. "Atmospheric Methane and Nitrous Oxide of the Late Pleistocene from Antarctic Ice Cores." *Science* 310 (2005): 1,317–21. (N)

Stainforth, D. A., et al. "Uncertainty in Predictions of the Climate Response to Rising Levels of Greenhouse Gases." *Nature* 433 (2005): 403–6.

Steig, E. "Climate May Not Be Linked with Circulation Slowdown." *Nature* 439 (2006): doi: 10/1038/439660a. (N)

Stoll, H. "Climate Change: The Arctic Tells its Story." *Nature* 441 (2006): 579–581. (W)

Sugi, M., A. Noda, and N. Sato: "Influence of the Global Warming on Tropical Cyclone Climatology: An Experiment with the JMA Global Model." *Journal of the Meteorological Society of Japan* 80 (2002): 249–72.

Suroweicki, J. *The Wisdom of Crowds.* New York: Anchor Books 2005.

Tedesco, M. "A New Record in 2007 for Melting in Greenland." *Eos: Transactions of the American Geophysical Union* 88 (2007): 383.

Thompson, L. G., et al. "Kilimanjaro Ice Core Records: Evidence of Holocene Climate Change in Tropical Africa." *Science* 298 (2002): 589–93.

Trenberth, K., and J. W. Hurrell. "Spurious Trends in Satellite MSU Temperatures from Merging Different Satellite Records." *Nature* 397 (1997): 164–67.

Treydte, K. S., et al. "The Twentieth Century Was the Wettest Period in Northern Pakistan over the Past Millennium." *Nature* 440 (2006): 1,179–82. (W)

Trimble, S. W. "The Double Standard in Environmental Science." *Regulation* 30 (2007): 16–22.

Tunved, P., et al. "High Natural Aerosol Loading over Boreal Forests." *Science* 312 (2006): 261–63. (B)

Turner, J., et al. "Significant Warming of the Antarctic Winter Troposphere." *Science* 311 (2006): 1,914–17. (W)

Unisys Weather, 2008. http://weather.unisys.com/hurricane/atlantic/index.html.

University of Alabama-Huntsville (satellite temperatures). http://vortex.nsstc.uah.edu/data/msu/t2lt/tltglhmam_5.2.

University of East Anglia (temperature data). http://www.cru.uea.ac.uk/cru/data/temperature/.

U.S. Department of Agriculture, 2008. http://www.nass.usda.gov/QuickStats/.

Vecchi, G. A., et al. "Weakening of Tropical Pacific Atmospheric Circulation due to Anthropogenic Forcing." *Nature* 441 (2006): 73–76. (W)

Velicogna, I., and J. Wahr. "Measurements of Time-Variable Gravity Show Mass Loss in Antarctica." *Science* 311 (2006): 1,754–56.

Wadhams, P., and W. Munk. "Ocean Freshening, Sea Level Rising, Sea Ice Melting." *Geophysical Research Letters* 31 (2004): doi:10.1029/2004GL020039.

Walker, G. "Climate Change: The Tipping Point of the Iceberg." *Nature* 441 (2006): 802–05. (W)

Wara, M. W., A. C. Ravelo, and M. L. Delaney. "Permanent El Niño–Like Conditions during the Pliocene Warm Period." *Science* 309 (2005): 758–61. (W)

Weatherhead, E. C., and S. B. Andersen. "The Search for Signs of Recovery of the Ozone Layer." *Nature* 441 (2006): 39–45. (W)

Webb, T., III, et al. "Late Quaternary Climate Change in Eastern North America: A Comparison of Pollen-Derived Estimates with Climate Model Results." *Quaternary Science Reviews* 16 (1998): 587–606at (available at http://climate.gi.alaska.edu/ClimTrends/change/4903Change.html).

Webster, P. J., et al. "Changes in Tropical Cyclone Number, Duration, and Intensity in a Warming Environment." *Science* 309 (2005): 1,844–46. (W)

Wegman, E. *Controversy in Global Warming: A Case Study in Statistics.* Wiley-Interscience, 2007.

Weidick, A. *Satellite Image Atlas of Glaciers of the World: Greenland.* U.S. Geological Survey Professional Paper 1386-C. Washington: U.S. Government Printing Office, 1995.

Wentz, F. J., and M. Schabel. "Effects of Orbital Decay on Satellite-Derived Lower-Tropospheric Temperatuer Trends." *Nature* 394 (1998): 661–64.

Witze, A. "Meteorology: Bad Weather Ahead." *Nature* 441 (2006): 564–66. (W)

Wöppelmann, G., et al. "Geocentric Sea-Level Trend Estimates from GPS Analyses at Relevant Tide Gauges Worldwide." *Global and Planetary Change* 57 (2007): 396–406.

Yoshimura, J., M. Sugi, and A. Noda. "Influence of Greenhouse Warming on Tropical Cyclone Frequency." *Journal of the Meteorological Society of Japan* 84 (2006): 405–28.

Zimov, S. A., E. A. G. Schuur, and F. S. Chapin III. "Permafrost and the Global Carbon Budget." *Science* 312 (2006): 1,612–13. (W)

Index

Articles, papers, and reports are indexed by primary author(s) cited in the text and under the source of the article by title (if given) or topic.

About the Authors

Patrick J. Michaels is senior fellow in environmental studies at the Cato Institute and served as a research professor of environmental sciences at the University of Virginia for 30 years. He is a member of the Intergovernmental Panel on Climate Change, which shared the 2007 Nobel Peace Prize; past president of the American Association of State Climatologists; and winner of the American Library Association's worldwide competition for public service writing. In a field dominated by sensationalist media coverage and disputed research, Michaels has earned a national reputation for providing rigorous and clearly explained analyses of the science and public policy of global warming. One of the most prolific and quoted experts in the field, he is the author of five books, including *Meltdown: The Predictable Distortion of Global Warming by Scientists, Politicians, and the Media*, and (co-authored with Robert Balling Jr.) *The Satanic Gases: Clearing the Air about Global Warming*. In addition, he has written hundreds of popular, technical, and scientific articles on climate change and society. His only skepticism is his conclusion that the world is not going to come to an end from global warming.

Robert C. Balling Jr. is a professor in the climatology program in the School of Geographical Sciences at Arizona State University. He has published over 125 articles in the professional scientific literature, has been the recipient of a numerous research grants, lectured throughout the United States and more than a dozen foreign countries, and has appeared in a number of scientific documentaries and news features. He has served as a climate consultant to the United Nations Environment Program, the World Climate Program, the World Meteorological Organization, and the Intergovernmental Panel on Climate Change. Balling's books on climate change include *The Heated Debate: Greenhouse Predictions Versus Climate Reality; Interactions of Desertification and Climate* (published by the United Nations); *and The Satanic Gases: Clearing the Air about Global Warming*, which he co-authored with Patrick J. Michaels.

Cato Institute

Founded in 1977, the Cato Institute is a public policy research foundation dedicated to broadening the parameters of policy debate to allow consideration of more options that are consistent with the traditional American principles of limited government, individual liberty, and peace. To that end, the Institute strives to achieve greater involvement of the intelligent, concerned lay public in questions of policy and the proper role of government.

The Institute is named for *Cato's Letters,* libertarian pamphlets that were widely read in the American Colonies in the early 18th century and played a major role in laying the philosophical foundation for the American Revolution.

Despite the achievement of the nation's Founders, today virtually no aspect of life is free from government encroachment. A pervasive intolerance for individual rights is shown by government's arbitrary intrusions into private economic transactions and its disregard for civil liberties.

To counter that trend, the Cato Institute undertakes an extensive publications program that addresses the complete spectrum of policy issues. Books, monographs, and shorter studies are commissioned to examine federal budget, Social Security, regulation, military spending, international trade, and myriad other issues. Major policy conferences are held throughout the year, from which papers are published thrice yearly in the *Cato Journal.* The Institute also publishes the quarterly magazine *Regulation.*

In order to maintain its independence, the Cato Institute accepts no government funding. Contributions are received from foundations, corporations, and individuals, and other revenue is generated from the sale of publications. The Institute is a nonprofit, tax-exempt, educational foundation under Section 501(c)3 of the Internal Revenue Code.

CATO INSTITUTE
1000 Massachusetts Ave., N.W.
Washington, D.C. 20001
www.cato.org